Bridge Optimization - Inspection and Condition Monitoring

Edited by Yun Lai Zhou and Magd Abdel Wahab

Published in London, United Kingdom

IntechOpen

Supporting open minds since 2005

Bridge Optimization - Inspection and Condition Monitoring
http://dx.doi.org/10.5772/intechopen.79062
Edited by Yun Lai Zhou and Magd Abdel Wahab

Contributors
Ali Asghar Bataleblu, Yun Lai Zhou, Chung Fu, Gengwen Zhao, Yang Lu, Tim Saad, Mitsuru Uesaka, Yuki Mitsuya, Katsuhiro Dobashi, Yoiichi Kusano, Eiji Yoshida, Yoshinobu Ooshima, Yasuhiro Ishida

First published in London, United Kingdom, 2020 by IntechOpen
IntechOpen is the global imprint of INTECHOPEN LIMITED, registered in England and Wales, registration number: 11086078, 7th floor, 10 Lower Thames Street, London, EC3R 6AF, United Kingdom
Printed in Croatia

British Library Cataloguing-in-Publication Data
A catalogue record for this book is available from the British Library

Additional hard and PDF copies can be obtained from orders@intechopen.com

Bridge Optimization - Inspection and Condition Monitoring
Edited by Yun Lai Zhou and Magd Abdel Wahab
p. cm.
Print ISBN 978-1-78985-260-8
Online ISBN 978-1-78985-271-4
eBook (PDF) ISBN 978-1-78985-272-1

For EU product safety concerns:
IN TECH d.o.o., Prolaz Marije Krucifikse Kozulić 3, 51000 Rijeka, Croatia, info@intechopen.com or visit our website at intechopen.com.

Meet the editors

Dr. Yun Lai Zhou is a research fellow in the Department of Civil and Environmental Engineering at Hong Kong Polytechnic University, China. He received his BSc in 2010 in Theoretical and Applied Mechanics from Northwestern Polytechnic University, China, and his PhD in 2015 in Civil Engineering from the Technical University of Madrid, Spain. Dr. Zhou has published more than 50 scientific papers, conference proceedings, and technical reports in structural health monitoring, transmissibility-based damage identification, machine learning, and structural dynamics. His current research interests include structural health monitoring, sensing techniques, data processing, fracture mechanics, damage mechanics, coupling, and structural vibrations.

Prof. Magd Abdel Wahab is a Professor of Applied Mechanics in the Department of Mechanical Construction and Production at Ghent University, Belgium. He received his BSc, 1988, in Civil Engineering and his MSc, 1991, in Structural Mechanics, both from Cairo University. Prof Wahab completed his PhD in Fracture Mechanics in 1995 at KU Leuven, Belgium. He was awarded a Doctor of Science degree from the University of Surrey in 2008. He has published more than 240 scientific papers and technical reports in solid mechanics and dynamics of structures. His research interests include fracture mechanics, damage mechanics, fatigue of materials, durability, and dynamics and vibration of structures.

Contents

Preface

This book aims to address finite element analysis, sensing and inspection techniques, and applications for bridge condition monitoring and damage identification. It mainly focuses on finite element analysis, sensory systems, and signal-processing methodologies, enabling the determination of defects from response analysis and also the assessment of bridge condition.

Bridge condition monitoring is a complex task, involving multidisciplinary contributions, various techniques, and methodologies that have been developed over the past few decades. These include sensing techniques such as X-ray testing, ultrasonic testing, optical fiber sensing, piezoelectric transduction, and magnetic particle testing.

Each chapter has been written by specialists in that field, with rich experience in practical engineering. This book will provide a useful reference to beginners in this field, and the diverse introduction of techniques and approaches will also be invaluable for experienced researchers and engineers.

Dr. Yun Lai Zhou
Department of Civil and Environmental Engineering
at Hong Kong Polytechnic University,
China

Magd Abdel Wahab
Professor,
Ghent University,
Ghent, Belgium

Section 1

Introduction

Chapter 1

Introductory Chapter: Some Insights into Bridge Structural Condition Monitoring

Yun-Lai Zhou and Linya Liu

1. Introduction

Bridge structural condition monitoring has become a hot spot in both research and engineering fields. Bridges, serving as a connection between cliffs, shallow rivers, or special environmental conditions, have a lot of forms in functionality, economy, and art consideration. For instance, concrete bridge, steel bridges, cable stayed bridge, suspension bridges shown in **Figure 1,** and so on have been served in various cities [1]. The initial use of bridge is for functionality like footbridge [2], and then other considerations are included. As shown in **Figure 1**, both the suspension bridge and cable-stayed bridge are extending for long span and large area application in civil engineering.

Since the bridges provide the convenient transportation for passengers and vehicles, the in-service safety shall be the most essential issue in the lifecycle service of bridges, providing timely early stage warning and suggestion for the possible maintenance [3–7]. For instance, fiber optic sensors are applied for full-scale destructive bridge condition monitoring [3]. A comprehensive discussion on bridge instrumentation and monitoring for structural diagnosis is conducted in [4], expressing the general steps for bridge management system for condition monitoring including experimental tests, nondestructive tests, performance evaluation, and so on; in [5], the temperature effect is studied between the temperature and the frequency ratio for model plate-girder bridges under uncertain temperature conditions; and also the modal strain energy is extended to predict the location and severity of the damages; in [6], the acoustic emission is applied for monitoring the prestressed concrete bridges health condition after constructing a reference-signals database; in [7], Poisson process is applied to simulate the arrival of vehicles traversing a bridge, and a stochastic model of traffic excitation on bridges is constructed to be incorporated in a Bayesian framework, to assess the properties and update the uncertainty for condition evaluation of the bridge superstructure.

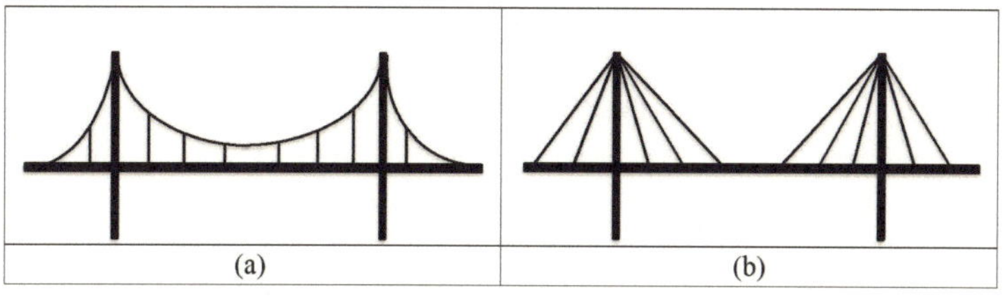

Figure 1.
Diagram for (a) suspension bridge; (b) cable-stayed bridge.

IntechOpen

1.1 Sensing techniques

Sensing techniques include a lot of conventional technologies and advanced technologies developed in recent decades. The conventional technologies include impact echo testing, dye penetration, strain gage, electrical magnetic testing, piezoelectric gages, acoustic emission, leakage testing, magnetic testing, ultrasonic testing, radiographic testing, eddy current testing, infrared thermography testing, microwave testing, and so on. Advanced technologies include phased array ultrasonic testing, eddy current array testing, microelectromechanical sensors, air-coupled sensors, vision sensors with cameras, radar sensors, and so on. In [8], the condition monitoring system for Tsing Ma Bridge is thoroughly introduced: the wind and structural health monitoring (WASHMS) in the Tsing Ma Bridge used about 300 sensors: anemometers, temperature sensors, accelerometers, strain gauges, displacement transducers, weigh-in-motion sensors, and so on. The Global Positioning System-On-Structure Instrumentation System (GPS-OSIS) was installed to improve the bridge displacement response monitoring. In [9], the Global Navigation Satellite System (GNSS): BeiDou Navigation Satellite System (BDS) and Global Positioning System (GPS) are applied to monitoring the bridge displacement responses. In [10], image processing is applied to construct a vision-based monitoring system for cable tension estimation under various weather conditions in the cable-stayed bridge, proving that the natural frequencies can be obtained up to the third and fifth modes. In [11], the radar sensor techniques were employed to predict the changes in the natural frequencies of bridge girders with certain characteristics that control the structural performance with being incorporated with computational modeling. In [12], commercially available remote sensors for Highway Bridge condition evaluation such as ground penetrating radar (GPR), optical interferometry, digital image processing (DIC), and so on are summarized.

1.2 Damage identification

Damage identification, part of structural health monitoring (SHM), has appealed lots of attention since the occurrence of defect/damage demands repairing and maintenance, simultaneously leading to economic loss. Damage identification techniques can be summarized into two categories: model based and data based, and this has been discussed in [1].

Modal testing serves as the fundamental and most essential technique in SHM, and along with the development of technology, modal testing underwent experimental modal analysis (EMA) and operational modal analysis (OMA). The key difference between them is EMA needs, while OMA does not demand the measurement of excitation. During the last decades, new measurement techniques also arise in engineering application. For instance, ultrasonic testing, eddy current, magnetic particle testing, acoustic emission, and so on are all imposed in SHM.

Comparing with the model-based techniques, data based, especially the output only based damage identification suggests a wide applicable potential since its merit relying on structural responses solely. Transmissibility is a typical output only based technique [13], which has been developed in the past decades for system identification, damage detection, localization, quantification, and assessment [14]. Review can refer to [14]. Even a lot of investigations about transmissibility can be accessed to [15–19], for instance, transmissibility coherence (TC) is raised [14, 17]; cosine-based indicator is constructed from the modal assurance criterion (MAC) and incorporated with transmissibility for damage detection and quantification relatively [15]; Mahalanobis distance is also applied to transmissibility for damage detection [16]; transmissibility is extended to apply in the responses analysis of ultrasonic testing [19]; the transmissibility still encounters difficulty in both theory development and engineering application. This study tries to extend the transmissibility theory for estimating and reconstructing mode shape from structural responses solely. Remaining work can be summarized as follows: Section 2 gives the theoretical development of transmissibility mode shape (TMS) and comparison between transmissibility-based OMA and frequency response functions (FRFs) based EMA, Section 3 gives the possible damage indicators, and Section 4 gives the numerical case study; conclusions are finally summarized.

2. Structural condition monitoring

2.1 Transmissibility and transmissibility coherence

Transmissibility has several kinds of definitions with existing reviews [14], while the fundamental concept is the ratio between two structural responses, which can be expressed as

$$T_{(i,s)} = \frac{X_i}{X_s} \tag{1}$$

where i, s mean the response locations, while X_i and X_j represent the frequency spectrum of dynamic response x_i and x_s in time domain.

Transmissibility can be assessed with several ways, for instance, to use FRFs if available,

$$T_{(i,s)} = \frac{X_i}{X_s} = \frac{H_{ir}}{H_{sr}} \tag{2}$$

where r denotes the excitation location (assuming single load). H represents the FRFs.

Similar to the application of coherence in FRFs analysis, TC is also raised and defined as

$$TC_{(i,s)} = \left| \frac{G_{is}^2}{G_{ii}G_{ss}} \right| \tag{3}$$

where G means the auto- and cross spectrum. TC is initially developed for damage/small nonlinearity detection and quantification, and later it is advanced for natural frequency extraction.

2.2 Transmissibility mode shape (TMS)

For a single load linear elastic structural system, the *FRF* can also be expressed as

$$H_{(i,r)}(\omega) = \sum_{p=1}^{n} \frac{\phi_p^i \phi_p^r}{k_p - \omega^2 m_p + j\omega c_p} \tag{4}$$

where p denotes the pth mode, n means the number of modes considered. k_p, m_p, and c_p mean modal stiffness, mass, and damping, respectively, ϕ means the mode shape, and ω represents the frequency.

Then, the transmissibility illustrated above can be further expressed as

$$T_{(i,s)} = \frac{X_i}{X_s} = \frac{H_{ir}}{H_{sr}} = \frac{\sum_{p=1}^{n} \frac{\phi_p^i \phi_p^r}{k_p - \omega^2 m_p + j\omega c_p}}{\sum_{p=1}^{n} \frac{\phi_p^s \phi_p^r}{k_p - \omega^2 m_p + j\omega c_p}} = \frac{\phi^r}{\phi^s} \tag{5}$$

Note, this relation shall be better if obtained by using Laplace transform. As to Eq. (5), if fixing s, transmissibility will allow assessing the mode shape, or scalar mode shape. For each mode, transmissibility will express the scalar mode shape at each location, if further obtaining the direction of the scalar mode shape in each measurement location, and then the transmissibility-based mode shape TMS (full-unscaled mode shape ϕ_{1s}, ϕ_{2s}, ϕ_{3s}, …, ϕ_{Ns}, N is the number of measured responses) will be obtained. A general definition can be denoted as

$$TMS\Big|_{(i,p)}^{B} = \int_{f_{B1}}^{f_{B2}} T_{(i,s)} df \tag{6}$$

where B denotes the frequency boundary [B1, B2] around the natural frequency, and f indicates the frequency domain. TMS_p means p^{th} TMS. All the TMSs will later contribute for further OMA [14].

2.3 Comparison between transmissibility and FRF

Table 1 illustrates the comparison between EMA and transmissibility-based OMA, and it can be found that transmissibility has been developed by analog of FRF, where transmissibility can also perform the same function as FRF, like in damage detection, system identification, and so on. Note that transmissibility has not been thoroughly investigated; and further study is still needed to unveil new features.

Since transmissibility can assess TMS and natural frequencies, then, the extended parameters based on modal parameters can later similarly be applied in transmissibility-based OMA.

2.4 Transmissibility application for outlier identification

Damage identification includes several stages: detection, locating, quantification, and remaining life assessment. All damage identifications follow the same procedure, (1) operational evaluation; (2) data acquisition, fusion, and cleansing; (3) feature extraction and information condensation; and (4) statistical mode development for feature discrimination [20]. The most essential step is feature extraction. Generally, feature means the property associated with the structural

Modal analysis	EMA	Transmissibility-based OMA
Kernel	FRF	Transmissibility
Coherence	FRF coherence	TC [17]
Modal parameters	Mode shape	TMS
	Frequency extraction techniques like SSI	Frequency extraction techniques like TC based [14]

Table 1.
Comparison between EMA and transmissibility-based OMA.

internal change. For instance, the cross section reduction will result in stiffness reduction, which later changes the structural dynamic responses. Then, features can be constructed from structural dynamic responses to assess the stiffness reduction (kind of damage).

Certainly, damage has more kinds, like spalling in concrete structures, corrosion induced defects, and so on. These kinds of defects at initial stage may not cause a clear change in stiffness reduction; thus, special techniques like acoustic emission should be adopted in further investigation.

For the damage illustrated in this study-stiffness reduction related damage, vibration-based techniques are taken into consideration. To construct damage indicator, the change of feature is the commonest, and one may use MAC for achieving a comparable indicator without needing normalization, which can be denoted as

$$DI = 1 - MAC = 1 - \frac{\left((TMS^u)^T \times (TMS^d)\right)^2}{\left((TMS^u)^T \times (TMS^u)\right) \times \left((TMS^d)^T \times (TMS^d)\right)} \tag{7}$$

where $(TMS)^u$ and $(TMS)^d$ denote the value under undamaged and damaged states, respectively.

Of course, herein, more indicators can be constructed, since TMS and natural frequencies are assessed by transmissibility in OMA, curvature, higher order derivative, and so on, and all these modal parameters based indicators can further be applied in damage detection [23].

3. Case study

In order to illustrate the feasibility of the proposed methodology, a pined-pined beam is numerically analyzed. Young's modulus is 185.2 GPa, dimension is $0.005 \times 0.006 \times 1.000$ m, density is 7800 Kg/m^3, and a vertical impulse is excited in the node 7 with 10 elements discretized on average in the whole beam. Dynamic responses are considered in the further OMA. The schematic diagram in **Figure 2** shows the beam. Different damage levels are simulated with reducing the stiffness in element 3, and for damage level D1, D2, D3, and D4, the stiffness reduced from 5, 10, 15, to 20% accordingly.

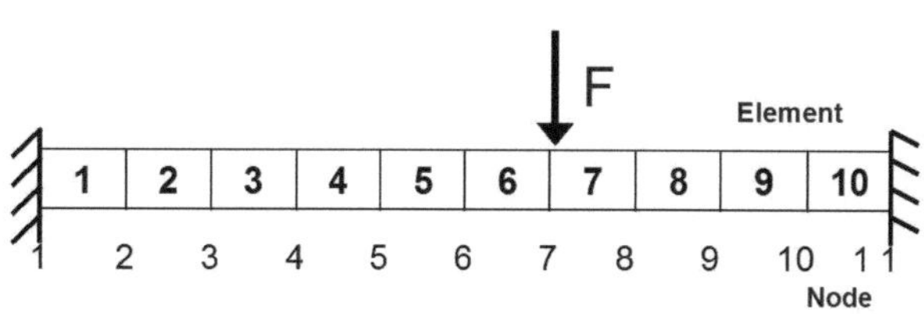

Figure 2.
Schematic diagram of the pined-pined beam.

4. Results and discussion

Results for the aforementioned methodology is computed and discussed in this section. **Figure 3** illustrates the mode shapes for first four modes, while **Figure 4** demonstrates the TMSs for first four modes, where one can find that both mode shapes and TMSs share the similar shapes, and their well agreement implies the potential use of TMS in damage identification.

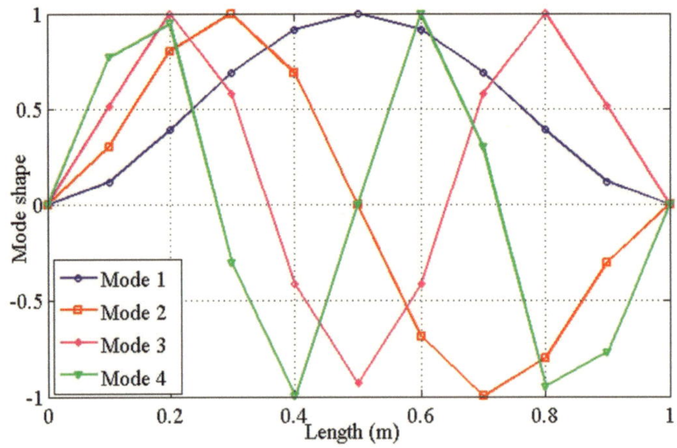

Figure 3.
Mode shapes of the beam for first four modes.

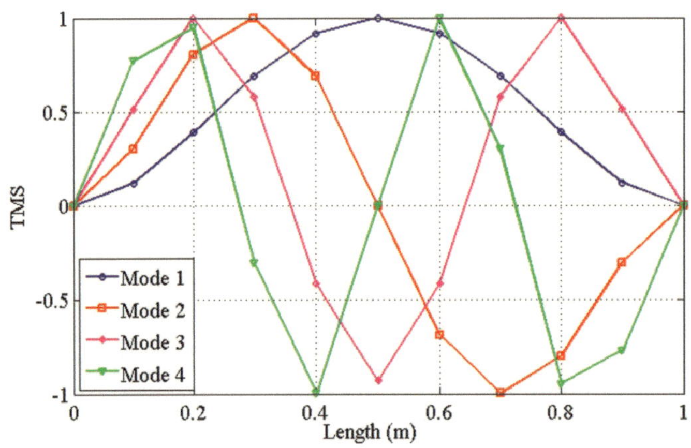

Figure 4.
TMSs of the beam for first four modes.

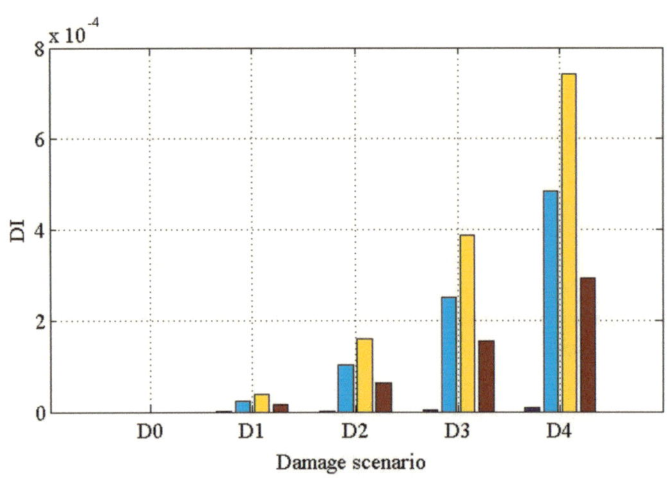

Figure 5.
DI for first four modes.

Figure 5 gives the structural detection results from the constructed damage indicator DI. From this figure, it can be found that all damage levels are detected. It should also be noted that the change of DI for all the four modes are not very much, this suggests that TMSs vary small before and after the occurrence of damage, and further enhancement should be conducted in order to achieve a better damage detection performance.

5. Concluding remarks

This study tries to discuss some insights for bridge condition monitoring, also extends the mode shape into transmissibility-based OMA, and by using transmissibility, TMS is assessed by analog of mode shape in EMA, which paves the way for further investigation of extending the mode shape-based indicators to TMS-based analysis [21, 22]. MAC is used to construct a damage indicator with being verified by a pined-pined beam. The damage detection performance implies further necessary investigation for obtaining a better and deeper understanding.

Author details

Yun-Lai Zhou[1]* and Linya Liu[2]

1 Department of Civil and Environmental Engineering, National University of Singapore, Singapore

2 Engineering Research Center of Railway Environment Vibration and Noise, Ministry of Education, East China Jiaotong University, Nanchang, China

*Address all correspondence to: zhouyunlai168168@gmail.com

References

[1] Cao H, Qian X, Zhou Y, Chen Z, Zhu H. Feasible range for midtower lateral stiffness in three-tower suspension bridges. Journal of Bridge Engineering. 2018;**23**(3):06017009

[2] Qin S, Zhou Y, Kang J. Footbridge serviceability analysis: From system identification to tuned mass damper implementation. KSCE Journal of Civil Engineering. 2019;**23**(2):754-762

[3] Storoy H, Sether J, Johannessen K. Fiber optical condition monitoring during a full scale destructive bridge test. Journal of Intelligent Material Systems and Structures. 1997;**8**:633-643

[4] Farhey DN. Bridge instrumentation and monitoring for structural diagnostics. Structural Health Monitoring. 2005;**4**(4):0301-0318

[5] Kim J-T, Park J-H, Lee B-J. Vibration-based damage monitoring in model plate-girder bridges under uncertain temperature conditions. Engineering Structures. 2007;**29**:1354-1365

[6] Kalicka M. Acoustic emission as a monitoring method in prestressed concrete bridges health condition evaluation. Journal of Acoustic Emission. 2009;**27**:18-26

[7] Chen Y, Feng MQ, Tan C-A. Bridge structural condition assessment based on vibration and traffic monitoring. Journal of Engineering Mechanics. 2009;**135**(8):747-758

[8] Xu Y-L. Making good use of structural health monitoring systems of long-span cable-supported bridges. Journal of Civil Structural Health Monitoring. 2018;**8**:477-497

[9] Xi R, Jiang W, Meng X, Chen H, Chen Q. Bridge monitoring using BDS-RTK and GPS-RTK techniques. Measurement. 2018;**120**:128-139

[10] Kim S-W, Jeon B-G, Cheung J-H, Kim S-D, Park J-B. Stay cable tension estimation using a vision-based monitoring system under various weather conditions. Journal of Civil Structural Health Monitoring. 2017;7: 343-357

[11] Maizuar M, Zhang L, Miramini S, Mendis P, Thompson RG. Detecting structural damage to bridge girders using radar interferometry and computational modeling. Journal of Structural Control and Health Monitoring. 2017;**10**(24):1-6

[12] Vaghefi K, Oats RC, harris DK, Ahlborn TM, Broooks CN, Endsley KA, et al. Evaluation of commercially available remote sensors for highway bridge condition assessment. Journal of Bridge Engineering. 2012;**17**(6):886-895

[13] Zhou Y, Maia NMM, Sampaio RPC, Wahab MA. Structural damage detection using transmissibility together hierarchical clustering analysis and similarity measure. Structural Health Monitoring-An International Journal. 2017;**16**(6):711-731

[14] Zhou Y. Structural Health Monitoring by Using Transmissibility [PhD thesis]. Spain: Technical University of Madrid; 2015

[15] Zhou Y, Wahab MA. Cosine based and extended transmissibility damage indicators for structural damage detection. Engineering Structures. 2017; **141**:175-183

[16] Zhou Y, Figueiredo E, Maia NMM, Sampaio R, Perera R. Damage detection in structures using a transmissibility-based Mahalanobis distance. Structural Control and Health Monitoring. 2015;**22**: 1209-1222

[17] Zhou Y, Figueiredo E, Maia NMM, Perera R. Damage detection and

quantification using transmissibility coherence analysis. Shock and Vibration. 2015. Article ID 290714, 16 pages

[18] Zhou Y, Cao H, Liu Q, Wahab MA. Output-based structural damage detection by using correlation analysis together with transmissibility. Materials. 2017;**10**:866

[19] Zhou Y, Qian X, Birnie A, Zhao X. A reference free ultrasonic phased array to identify surface cracks in welded steel pipes based on transmissibility. International Journal of Pressure Vessels and Piping. 2018;**168**:66-78

[20] Sohn H, Farrar CR, Hemez FM, Shunk DD, Stinemates DW, Nadler BR, Czarnecki JJ. (2004). A Review of Structural Health Monitoring Literature: 1996–2001, los Alamos National Laboratory Report, LA-13976-MS, 2004

[21] Kerrouche A, Boyle WJO, Gebremichael Y, Sun T, Grattan KTV, Täljsten B, et al. Field tests of fibre bragg grating sensors incorporated into CFRP for railway bridge strengthening condition monitoring. Sensors and Actuators A. 2008;**148**:68-74

[22] Dissanayake PBR, Karunananda PAK. Reliability index for structural health monitoring of aging bridges. Structural Health Monitoring. 2008; 7(2):0175-0179

[23] Zhou Y, Maia NMM, Wahab MA. Damage detection using transmissibility compressed by principal component analysis enhanced with distance measure. Journal of Vibration and Control. 2018;**24**(10):2001-2019

Section 2

Application of Artificial Intelligence

Chapter 2

Computational Intelligence and Its Applications in Uncertainty-Based Design Optimization

Ali Asghar Bataleblu

Abstract

The large computational cost, the curse of dimensionality and the multidisciplinary nature are known as the main challenges in dealing with real-world engineering optimization problems. The consideration of inevitable uncertainties in such problems will exacerbate mentioned difficulties as much as possible. Therefore, the computational intelligence methods (also known as surrogate-models or metamodels, which are computationally cheaper approximations of the true expensive function) have been considered as powerful paradigms to overcome or at least to alleviate the mentioned issues over the last three decades. This chapter presents an extensive survey on surrogate-assisted optimization (SAO) methods. The main focus areas are the working styles of surrogate-models and the management of the metamodels during the optimization process. In addition, challenges and future trends of this field of study are introduced. Then, a comparison study will be carried out by employing a novel evolution control strategies (ECS) and recently developed efficient global optimization (EGO) method in the framework of uncertainty-based design optimization (UDO). To conclude, some open research questions in this area are discussed.

Keywords: computational intelligence, metamodeling, surrogate-assisted optimization, uncertainty-based design optimization

1. Introduction

Motivated by industrial demands and development of more powerful optimization techniques, the engineering design community has undergone a major transformation. They are continually seeking new optimization challenges and to solve increasingly more complicated real-world engineering problems in the shortest feasible time. In order to achieve the best solution in dealing with complex real-world engineering optimization, the classical optimization methods are weak in convergence. In solving these design problems, an evolutionary algorithm may require thousands of function evaluations in order to provide a satisfactory solution, whereas each evaluation requires hours of computer run-time. To overcome such difficulties, researchers have applied sampling-based learning methods such as artificial neural networks, radial basis functions, and polynomial model. These methods can 'learn' the problem behaviors and approximate the function value. These approximation models can speed up the function evaluation as well as estimation the function value with an acceptable accuracy. Also, they can improve optimization performance and provide a better final solution. However, the application of computational intelligence methods to expensive optimization problems is not straightforward. It is important to note that accuracy is the most important criterion for evaluating a metamodel, since metamodels with a low accuracy may lead to local optima, or may even fail to obtain a satisfactory solution. Nevertheless, the choice of the surrogate model is depending on the design problem [1].

Viana et al. [2] and Jin [3] have presented a survey study about metamodeling techniques and their application in the design and analysis of computer experiments. Moreover, they proposed the future work directions to handle more complex simulations. The metamodeling/surrogate-modeling techniques approximate the real model in the entire design space bound that could help to reduce the running time of a complex problem considerably [4]. The simulation-based design problems using metamodels is reviewed extensively in [5–8]. Furthermore, researchers at Boeing and Rice University have proposed a number of mathematical techniques for application of metamodels in optimization problems [2, 9]. They have introduced some software packages to expedite the design and optimization process by using metamodels. These packages include "Optimus" developed by Tzannetakis et al. [10] and "DAKOTA" developed by Adams et al. [11]. Some of the common metamodeling techniques are including response surface method (RSM), artificial neural networks (ANN), Kriging, radial basis functions (RBF), and support vector regression (SVR) [12].

In the recent years, the large number of research and literature indicates the importance of using metamodeling techniques in the optimization. Horng and Lin [13] have proposed an evolutionary algorithm optimizer using metamodels in the ordinal optimization framework. They have used their algorithm to solve a stochastic optimization problem with a huge discrete design space. Sóbester et al. [14] have presented a research on improving the accuracy of metamodels in engineering design problems. Gong et al. [15] have proposed a metamodel with the small computational cost for design by evolutionary optimization algorithms. In order to consider the low and high fidelity model's information and make a trade-off between accuracy and computational cost, Zhou et al. [16] have introduced an active learning strategy for application of metamodeling. Belyaev et al. [17] have presented a new tool namely GTApprox to generate medium-scale metamodels for industrial design. Sun et al. [18] have introduced a swarm optimization algorithm based on the surrogate models. Recently, a strategy for reducing the running time has presented by Sayyafzadeh [19] that is based on a self-adaptive metamodeling approach. Also, a number of metamodeling strategies that could be used for the uncertainty-based design optimization have reviewed by Chatterjee et al. [20].

Despite the recent advances in the design optimization tools, researchers are still trying to surmount some other issues such as curse of dimensionality, the numerical noise, and handling mixed discrete/continuous variables. Surrogate-assisted optimization (SAO) and the evolution control strategies (ECS) are two newly developed methods in this field of area [1]. Both of these strategies can be applied offline or online. The main difference between these two methods is the management of using metamodels instead of real models during the optimization process. In the SAO strategy, metamodels are substituted for real models directly, but in the ECS strategy, metamodels are substituted for real models in some of the optimizer design points. Furthermore, metamodels that are used offline are not updated while the optimization is ongoing, whereas online metamodels are adaptively updated during the optimization process and can progressively improve the accuracy of the metamodels [12]. One of the known SAO strategies is known as the efficient global optimization (EGO) that operates based on the approximation of responses using a Gaussian process (e.g. Kriging or SVR) [1].

In order to build globally accurate metamodels, offline SAO methods may require more sample points and they may be computationally expensive. Instead, the online SAO methods could train with fewer sample points. One of the weaknesses of online SAO methods is that the few numbers of sample points in the first iterations can lead to a poor predictive capability of the metamodel. Therefore, this can entice the optimizer into a local optimum or infeasible regions in the design space [1]. To surmount them all, researchers have presented some techniques to call real models and metamodels beside each other during the optimization. For instance, the real model can be used to correct the fitness value of some/all individuals in some generations of evolutionary optimization algorithms. This is known as the management of the metamodels or the evolution control and has been applied in many literatures [12]. However, the best time for calling the real model or the metamodel is still a major challenge for the metamodels' management.

This chapter is introduced for the wide field of research and can be applied by readerships who are interested in the development of computational intelligence techniques for nowadays' expensive optimization problems. Moreover, a novel ECS that benefits from managing the use of metamodels for increasing the optimization accuracy is proposed in this chapter. The performance benefits of this proposed strategy include decreasing the computational cost as well as providing a global or near-global optimum solution. It is important to note that this strategy can be applied for both deterministic and non-deterministic optimization problems, with any optimization algorithms [1].

This chapter is organized as follows. Section 2 introduces the metamodeling approximation. Section 3 presents the proposed ECS strategy. Section 4 presents applications of the proposed strategy to some mathematical benchmark problems, and numerical results are discussed in detail. Section 5 presents the research conclusions as well as some directions for the future research.

2. Meta-modeling

Extensive research on design and optimization of engineering problems using metamodeling techniques has been done. These research fields are including sampling, metamodeling, validation, management and application, and so on. Over the years it has become prove that metamodeling provides a decision criteria role for designers [7].

Metamodeling involves (a) choosing an experimental design for generating design points, (b) function evaluation of generated design points, and then (c) choosing a model to represent the data and fitting the model to the observed data (see **Figure 1**).

After building the metamodels from the available dataset, the accuracy of the models should carefully be checked. When the metamodel is found to have acceptable accuracy, it can be employed for considered design and optimization studies. The metamodel type that is suitable for the approximation could vary depending on the intended use or the underlying problem's physic and design space. Different datasets could be appropriate for building metamodels. The process of where pick out the design points in the design space, i.e. how to spread the design points within the complete design space, is called the design of experiments (DOE) [21].

There are several options for each of metamodeling steps as shown in **Figure 2**, and three predominant ones are highlighted. For example, response surface methodology usually employs central composite designs, second order polynomials, and

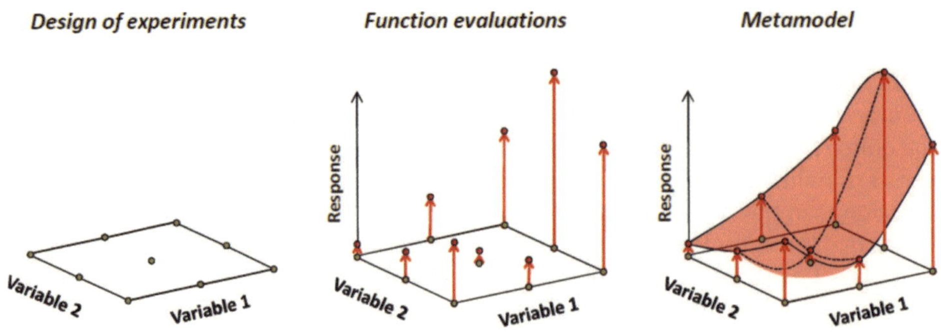

Figure 1.
The concept of metamodel creation [21].

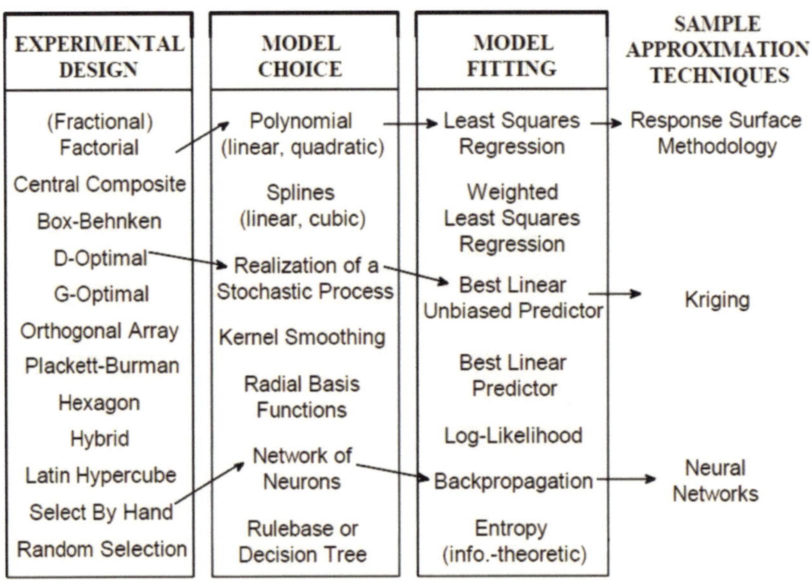

Figure 2.
Metamodeling techniques [5].

least squares regression analysis while building a neural network involves fitting a network of neurons by means of back-propagation to data which is typically hand selected [5].

Several strategies exist for finding the optimal solution using metamodel-based design optimization (MBDO). In what follows, a brief overview of several DOEs, metamodel choice and metamodel fitting, and some strategies of MBDO will be explained, respectively.

2.1 Design of experiments

The gathering a set of input-output date is the first step to build a metamodel and this dataset is known as the training set. The DOE is also the theory that helps to select best samples from the design space to cover everywhere. Based on the DOE theory, it is better that the training sets be space-filling and non-collapsing [1]. This signifies the importance of sampling efficiency in the generation of the training set for building an appropriate metamodel. This field has been a challenging research area among metamodeling researchers.

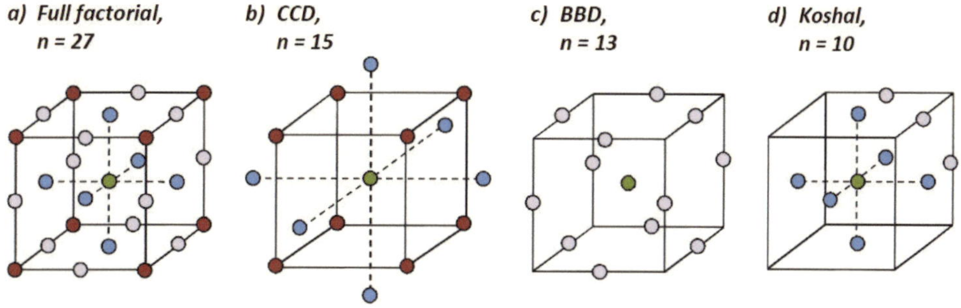

Figure 3.
Experimental designs in three variables for fitting second order models (Full factorial, central composite design (CCD), box-behnken design (BBD) and Koshal) [21].

2.1.1 Classical experimental designs

The idea of the classical DOE is to reach as much information as possible from a limited number of experiments. The focus of these methods is on planning the experiments so that the random error from the physical experiments has minimum influence in the approval or disapproval of a hypothesis. Therefore, a classical experimental design represents a sequence of experiments to be performed, expressed in terms of factors (design variables) set at specified levels (predefined values) [5].

Widely used "classical DOE" include factorial or fractional factorial designs (FFD), central composite designs (CCD), Box-Behnken designs (BBD), and Koshal designs (KD). Schematic illustration of these methods is presented in **Figure 3**. These classic methods tend to pick out the sample points around boundaries of the design space and leave a few at the center of the design space. To view the details of these methods, one may refer to [21].

2.1.2 Experimental designs for complex metamodels

As computer experiments involve mostly systematic error rather than random error as in physical experiments, researchers stated that in the presence of such sources of error, a good experimental design has to be space filling and non-collapsing rather than to concentrate on the boundary [5]. Also, in dealing with a complex design space, the metamodel's training samples should be spread the design points within the complete design space so that no prediction be too far from training points. Four types of space filling sampling methods that relatively more often used in the literature are orthogonal arrays, various Latin hypercube designs, Hammersley sequences, and uniform designs. Details of these methods are presented in Ref. [21].

2.2 Metamodel choice and metamodel fitting

After selecting an appropriate DOE strategy and performing the necessary computer runs, the next step is to choose a metamodel and fitting method. As alluded to earlier in the introduction, many machine learning methods such as ANN, Kriging, RBF and SVR have been used to approximate complex relations between a set of inputs and outputs, and can thus be used as a metamodel.

Despite the various metamodel types that have been introduced so far, which model is suitable for use? Different metamodels have their unique properties and consequently, there is no universal model that always is the best choice. Instead, the suitable metamodel depends on the problem at hand [21]. They are bound to their special domains, and thus no comparative studies have been conducted on them.

On the other hand, the performance of metamodels is depending on the problem to be addressed, and multiple criteria need to be considered. Model accuracy is probably the most important criterion, since approximate models with a low accuracy may lead the optimization process to local optima or even diverge from the optimal solution. Model accuracy also should be evaluated based on the new random sample points instead of the training data set points. The reason for this is that for some models overfitting is a common difficulty. In the case of overfitting, the model yields good accuracy on training data but may have poor performance on new sample points. The optimization process could easily go in the wrong direction if it is assisted by models with low accuracy [12].

There are some accuracy measures that may be used to evaluate the metamodels. The coefficient of determination R^2 is a measure of how well the metamodel is able to capture the variability in the dataset. Other common ways for accuracy measures include: the maximum absolute error (MAE), the average absolute error (AAE), the mean squared error (MSE) and the root mean squared error (RMSE) [21].

2.3 Metamodel-based design optimization

Metamodel-based design optimization can be applied using different strategies. The main issue with MBDO is the error that is introduced when approximating the real simulations with metamodels. The optimization process can be performed using the detailed simulation model, using its surrogate model, or both of them. Most common types of MBDO strategies are illustrated in **Figure 4**. In the first strategy (**Figure 4a**), a global metamodel will be built and then will be used during optimization. This approach uses a relatively large number of sample points at the outset and is commonly seen in the literature. The second strategy (**Figure 4b**) is based on the online metamodeling and involves the validation and/or optimization in the loop in deciding the resampling and remodeling strategy. In this strategy, samples will be generated iteratively to update the train data and related metamodel to maintain the model accuracy. In the third strategy (**Figure 4c**), the optimization is performed by adaptive sampling alone and no formal optimization process is used. This strategy directly generates new sample points toward the optimum with the guidance of a metamodel [7].

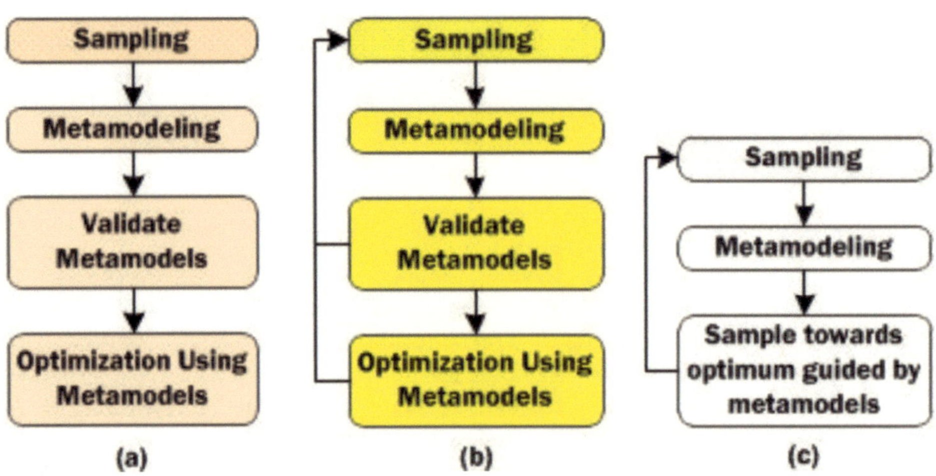

Figure 4.
MBDO strategies: (a) sequential approach; (b) adaptive MBDO; and (c) direct sampling approach [7].

In complex real-world design problems, achieving a flawless metamodel is almost impossible. Therefore, in order to take advantage of metamodels in MBDO, it is better to manage using metamodels based on their accuracy in design points/spaces. As shown in **Figure 5**, evolution control and migration are two major classes of management strategies for utilizing metamodels [12]. In the evolution control class, metamodels are called beside the real-models during the optimization process where the real-models are used in some/all individuals and in some/all generations. In model Migration, the entire population is divided into several sub-populations with its local metamodel. Also, the individuals in various sub-populations can migrate into other sub-populations [1]. To study the details of evolution control and migration, readers may refer to Tenne and Goh [12]. To improve the applicability of MBDO for complex real-world design problems, a novel ECS is developed that is introduced in the next section.

3. Proposed evolution control strategy (ECS)

In this section, a novel management strategy for application of the metamodels is introduced. This strategy relies on the Mean-Squared-Displacement (MSD) concept and is based on the evolution control class of metamodel management strategies. The MSD means the deviation of a particle's position relative to a reference position that is a statistical concept.

During the optimization process, the value of MSD for each design point must be computed that is named as calculated MSD (CMSD).

$$CMSD = \frac{1}{N_{train}} \sum_{n=1}^{N_{train}} (x_n - x_{ind})^2 \quad (1)$$

where N_{train} and X_n indicate the number and the vector of design variables of metamodel training data, respectively. X_{ind} is the vector of optimizer design variables, iteratively.

In order to use the proposed strategy in the optimization process, the MSD value of each sample that is used as metamodel's test point based on the all of training data set has to be computed. Then, using these MSD values, two MSD values for the first and last iteration of the optimization process have to be selected. These two values that are named initial MSD (IMSD) and final MSD (FMSD) respectively indicate

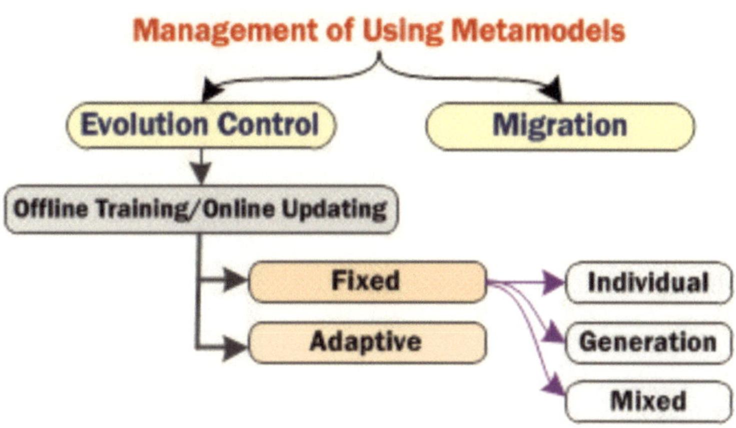

Figure 5.
Metamodel management strategies in MBDO [12].

the acceptable accuracy of metamodels from the first until the last iteration of the optimization process. An adaptive threshold namely predetermined MSD (PMSD) is proposed for the management of the decreasing PMSD value from IMSD to FMSD. The adaptive threshold of the PMSD enables the optimizer to call more metamodels vs. real models in the first iterations of optimization. Also, it enables that while the optimization is ongoing, the number of metamodel's call functions decrease slowly and the real model's call functions will increase. Here, the proposed adaptive PMSD threshold relies on the inverse hyperbolic cosecant concept, as follows:

$$PMSD = IMSD + (IMSD - FMSD) \times a\,\text{csc}\,h(k) \tag{2}$$

The variable k has to iteratively increase within an interval (e.g. from -12.5 to -1) while the optimization is ongoing. The introduced strategy is summarized iteratively, as follows [1]:

1. Calculate MSD value of metamodel's test points and initialize the IMSD and FMSD.

2. Determine the value of the k for PMSD estimation during the optimization process.

3. Start the optimization process using an initial design point.

4. Calculate the CMSD for each optimizer design point, compare CMSD value with PMSD and take a decision on using metamodels or real models.

5. Evaluate objective functions and constraints based on the decision in step 4.

6. Go to the next iteration of the optimizer and update the PMSD value.

7. Check the optimization convergence criterion and go to step 4.

The introduced strategy could be applied to all class of the optimization algorithms and both deterministic and non-deterministic optimization problems. In the next section, a number of benchmark problems are solved to present the ability of the proposed strategy.

4. MBDO of benchmark problems

In this section, the performance of the proposed strategy in achieving the global or at least the near-global optimum is investigated through solving some benchmark problems.

4.1 Analytical problem: one dimensional

Here, a one-dimensional nonlinear analytical example from Ref. [8] is used to illustrate the implementation of the proposed strategy. The mathematical formulation is shown as:

$$f(x) = (6x - 2)^2 sin(12x - 4) \tag{3}$$

The design variable x follows a normal distribution $x \sim N(x, \sigma_x^2)$, where $\sigma_x = 0.08$ and $x \in [0, 1]$. The objective of this example is to find the robust optimum solution and its cost function is defined as:

$$Minimize\ F(x) = \mu_f(x) + 3\sigma_f(x) \tag{4}$$

Based on four initial samples at x = [0, 0.33, 0.67, 1], a Kriging and ANN metamodel of Eq. (4) is constructed (**Figure 6**).

In **Figure 6**, the cross-mark and square mark represent the test and train sample points of metamodels, respectively. As illustrated in **Figure 6**, the robust optimum points resulted from Kriging and ANN metamodels are different from the real model one. The robust solution of Kriging and ANN are located at point x = 0.38, which are far away from the real solution x = 0.28. Therefore, due to the relatively large error of these metamodels, the obtained robust solution cannot be accepted. In order to resolve this issue, it is essential to add more samples to improve the prediction capability of metamodels.

Another way to overcome this problem is by using proposed ECS in this work. To do this, MSD value of all sample points that are used for metamodels test should be calculated. Then, based on these MSD values, a setting of the PMSD parameters (Eq. (2)) including IMSD and FMSD should be done. **Figure 7** illustrates the MSD value related to testing sample points. According to MSD values in **Figure 7**, the value of IMSD and FMSD are considered as 1.3 and 0.5, respectively. Now, it is time to select the adaptive threshold variable (k in Eq. (2)). This variable should increase iteratively while optimization is ongoing and its bound has a direct impact on how the PMSD threshold decreases adaptively. For example, by considering the interval [−12.5, −1] for the variable k and assuming the maximum iteration of the optimization process be 10, the PMSD adaptive threshold variation is illustrated in **Figure 8**.

Now, the optimization problem defined in this example is solved through proposed strategy along with simulated annealing optimizer and x = 1 as a start point. Convergence process in comparing with using only metamodels and real model is illustrated in **Figure 9**. Also, switching between real model and ANN metamodel

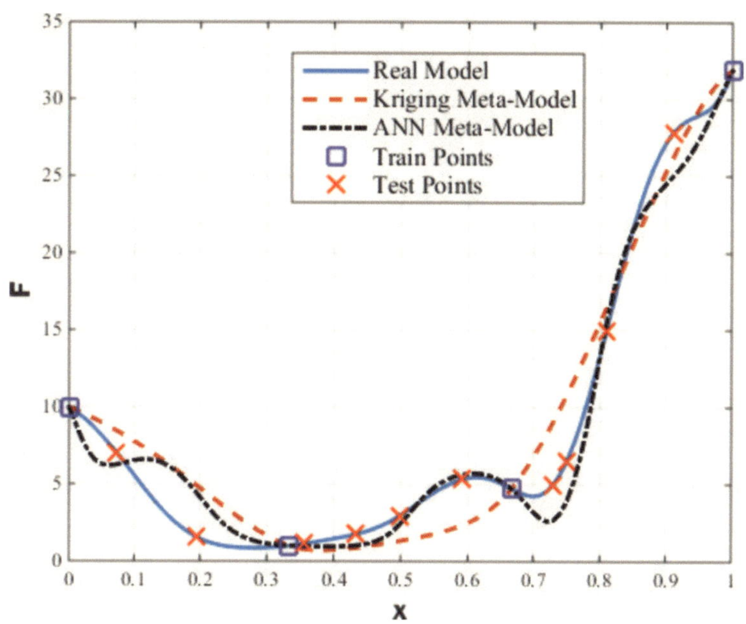

Figure 6.
Real robust function and its metamodels.

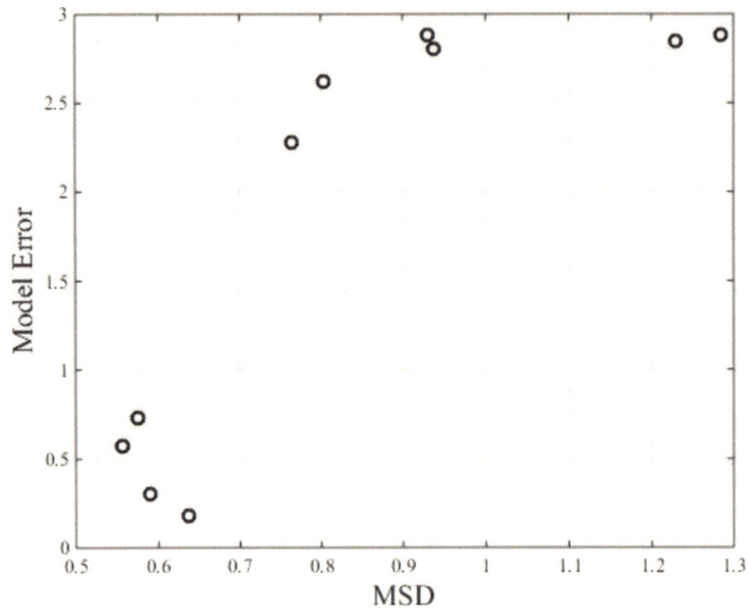

Figure 7.
MSD values related to the metamodels test points.

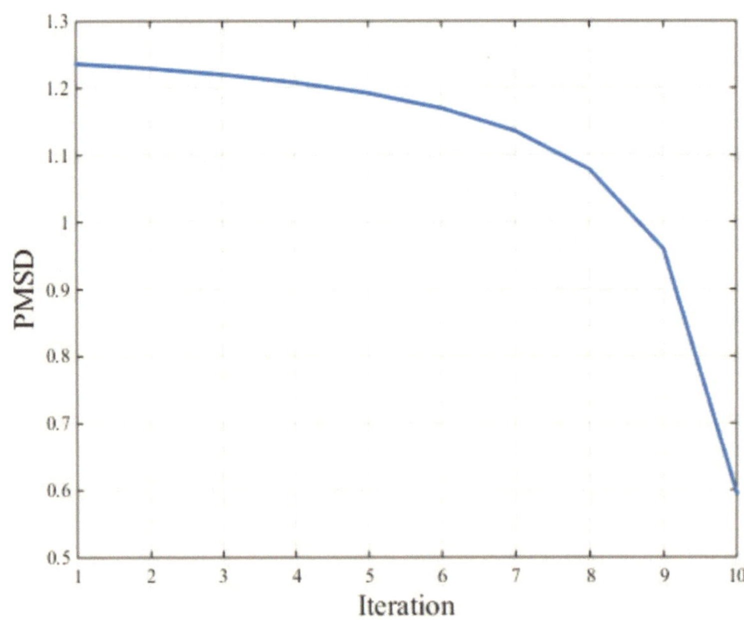

Figure 8.
PMSD variation vs. optimization iteration.

based on the CMSD value of each design point and PMSD value in the related iteration is shown in **Figure 10**. **Table 1** illustrates that proposed strategy with 3 real model call functions is achieved to near global robust optimum point compared to other methods.

In **Table 1**, the methods developed by Zhang et al. [22] that are based on the Kriging metamodel have been reached the near global optimal point through adding a new sample to the training set iteratively. For every new design point metamodel has been re-trained. Every time that a new point is added, you need to re-train and

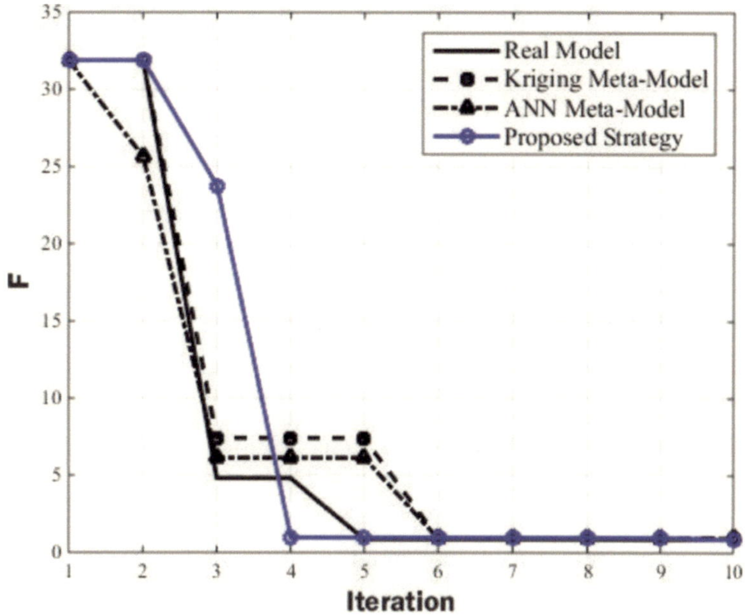

Figure 9.
Optimization convergence—one dimensional example.

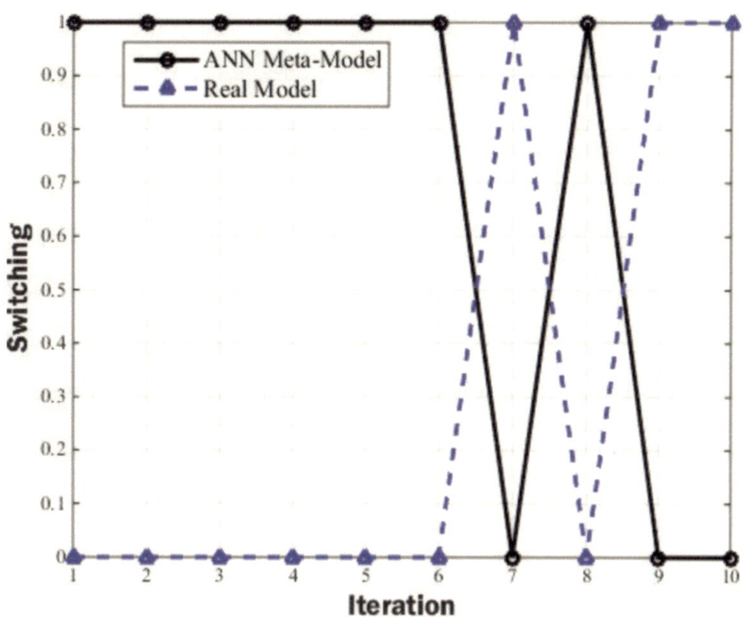

Figure 10.
Switching between ANN metamodel and real model.

re-test the metamodel. Since the metamodel is poor in the first iterations, this can mislead the optimization process into local optimum or infeasible regions in the design space.

As illustrated in **Figure 10**, the proposed strategy allows the optimizer to call metamodel in the first iterations. As optimization ongoing, metamodel accuracy in each design point will be checked and the real model will be called if necessary to prevent the optimizer from going to the wrong direction. Therefore, with proper

Model	Robust solution	
	X	F
Real model	0.3	0.88
Kriging	0.335	0.973
ANN	0.377	0.947
EI-based EGO [22]—(4 extra points is added to the training samples)	0.270	0.94
R-EI-based EGO [22]—(2 extra points is added to the training samples)	0.30	2.24
Proposed Strategy—(with 7 meta-model calls and 3 real model calls)	0.29	0.87

Table 1.
Robust solution resulted from different methods [1].

management of metamodels during the optimization process, the possibility of accessing the global or near global optimum will be increased.

4.2 Analytical problem—two dimensional

To further investigate the benefits of the proposed strategy, the robust design of two-dimensional Haupt function is presented here [22]. The Haupt function is defined as:

$$f(\boldsymbol{x}) = x_1 \sin(4x_1) + 1.1\, x_2 \sin(2\, x_2) \tag{5}$$

In this example, both of the design variables x_1 and x_2 follow a normal distribution $\mathbf{x} \sim N(\mathbf{x}, \sigma_{\mathbf{x}}^2)$, where $\mathbf{x} = [x_1, x_2]$ with $\mathbf{x} \in [0, 4]$ and $\sigma_{\mathbf{x}} = [\sigma_{x1}, \sigma_{x2}] = [0.2, 0.2]$. Considering the effect of design variable uncertainty, the robust design formulation is defined as:

$$Minimize\ F(\boldsymbol{x}) = \mu_f(x_1, x_2) + 3\sigma_f(x_1, x_2) \tag{6}$$

Based on Improved LHS (ILHS), 10 points are generated as the training sample points. The real model, Kriging and ANN metamodels of Eq. (6) are shown in **Figure 11**. It can be seen that the constructed metamodels are not sufficiently accurate and will mislead the optimizer into local optimum or non-optimal regions. To resolve this issue, Zhang et al. [22] have been proposed methods to generate new points while optimization is ongoing and increase the metamodel accuracy, iteratively. But since the predictive capability of the metamodel is poor in the first

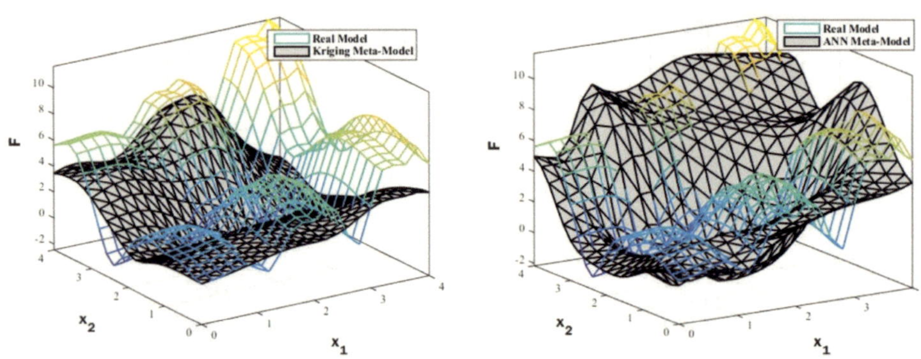

Figure 11.
Design space and different models of Haupt function.

iterations of optimization, there is not any guarantee to achieve global optimization, especially in complex real-world applications.

In order to the implementation of the proposed strategy to moderate this issue, PMSD equation parameters (Eq. (2)) including IMSD and FMSD should be determined based on the MSD amount of metamodels test points. Therefore, in this example, IMSD and FMSD values are considered as 2.4 and 0.8, respectively. As alluded to in the previous analytical example, to reduce the amount of PMSD threshold slowly, the interval of the variable k is assumed as $[-12.5, -1]$. According to the considered set of proposed strategy along with simulated annealing optimizer and x = [4, 4] as a start point, the robust design problem defined in Eq. (6) is solved using different methods. Optimization convergence process and switching between models are shown in **Figures 12** and **13**, respectively. Also, resulted robust optimums of different methods are presented in **Table 2**.

As presented in **Table 2**, one-stage sampling and sequential sampling methods that are based on the Kriging metamodel and proposed by Zhang et al. [22] have been reached the near global optimal point through 30 and 19 training sample points, respectively. But proposed strategy with checking the accuracy of the metamodel during optimization process through 5 switching between real model and metamodel (see **Figure 13**) is able to achieve near global optimum.

4.3 Engineering problem—two-bar truss structure

Uncertainty based design optimization of truss and frame structures is a popular topic in mechanical, civil, and structural engineering due to the complexity of problems and benefits to industry. In this section, the popular two-bar truss structure problem (**Figure 14**) is used as a benchmark problem for the multi-objective Robust Design Optimization (RDO) under epistemic uncertainties. The test case is adapted from Ref. [23].

As illustrated in **Figure 14**, the cross-section diameter (d) and the structure height (H) are as the design variables. The uncertain design parameters are including

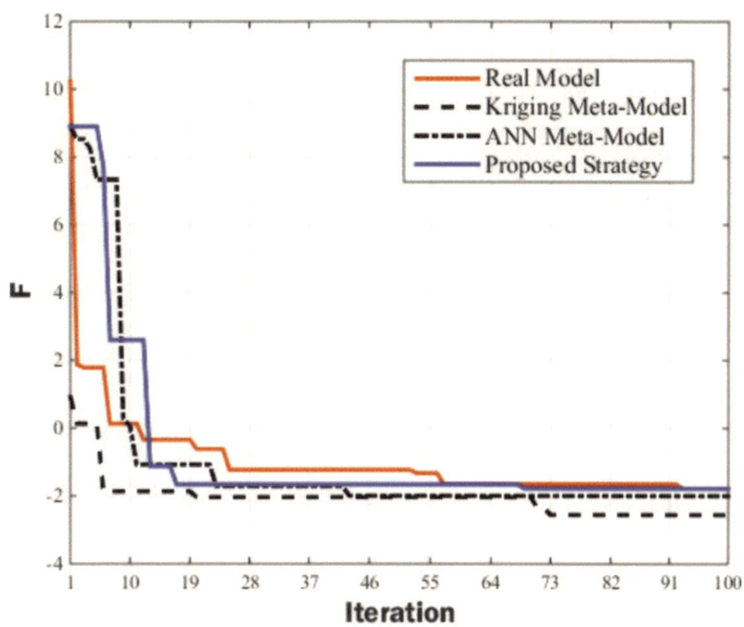

Figure 12.
Optimization convergence—Haupt function robust design.

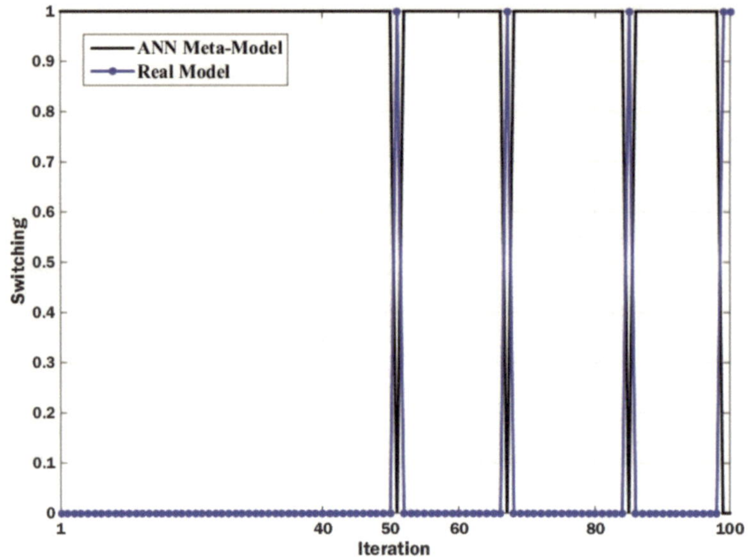

Figure 13.
Switching between metamodel and real model of Haupt function.

Model	Robust solution	
	X	F
Real model	[1.18, 2.45]	−1.78
Kriging	[2.72, 2.53]	−2.56
ANN	[1.74, 1.76]	−1.99
One-stage sampling method [22]—(30 training sample points)	[1.19, 2.44]	−1.74
sequential sampling method [22]—(19 training sample points)	[1.2, 2.47]	−1.68
Proposed strategy—(with 95 meta-model calls and 5 real model calls)	[1.2, 2.4]	−1.77

Table 2.
Robust solution resulted from different methods for Haupt function [1].

vertical force (P ~ N (150, 5) kN), structure width (B ~ N (750, 10) mm), Elastic modulus (E ~ N (2.1e5, 5e3) N/mm^2), and member thickness (t ~ N (2.5, 0.4) mm). The RDO problem is formulated in the following equation to minimize volume, vertical displacement and robustness criteria of the structure subject to constraints of stress and buckling.

In this design problem, the robustness measure F_{RC} given in Eq. (7) is defined as follows, with P, B, E and t as the four uncertain parameters.

$$\begin{aligned}
&Minimize \quad \left\{\mu_{volume}, \mu_{deflection}, F_{RC}\left(\sigma_{volume}, \sigma_{deflection}, \sigma_S\right)\right\}\\
&Subject\ to \quad g_1 : \mu_S \leq S_{\max}\\
&\qquad g_2 : \mu_S \leq S_{crit}\\
&With\ respect\ to \quad 20 \leq d\,(mm) \leq 80,\ 200 \leq H\,(mm) \leq 1000\\
&volume = 2\pi dt\sqrt{B^2 + H^2};\ deflection = \frac{P\left(B^2H^2\right)^{\frac{3}{2}}}{(2\pi EdH)^2}\\
&S = \frac{P\sqrt{B^2+H^2}}{2\pi tdH}, \quad S_{crit} = \frac{\pi^2 E\left(t^2 + d^2\right)}{8\left(B^2 + H^2\right)}, \quad S_{max} = 400\ MPa
\end{aligned} \tag{7}$$

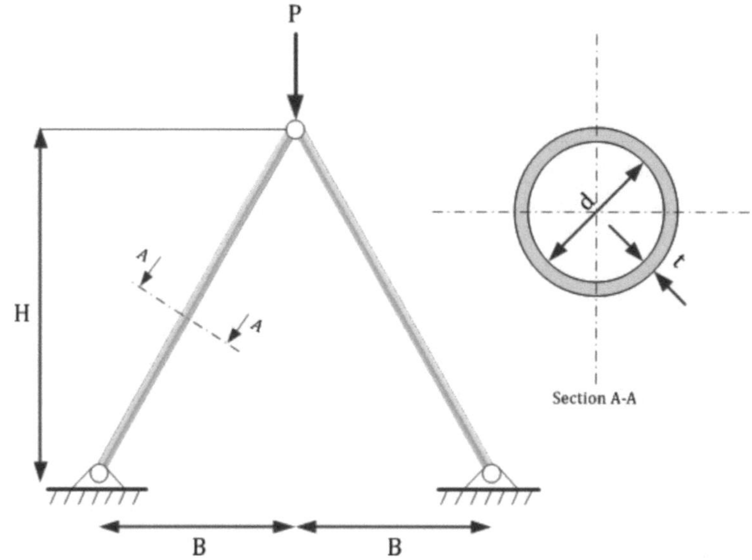

Figure 14.
Two-bar truss structure [23].

$$F_{RC} = \frac{1}{3 \times 4} \left\{ \begin{array}{l} \left(\frac{\sigma_{volume}}{\sigma_t} + \frac{\sigma_{volume}}{\sigma_w} \right) \\ + \left(\frac{\sigma_{deflection}}{\sigma_t} + \frac{\sigma_{deflection}}{\sigma_w} + \frac{\sigma_{deflection}}{\sigma_P} + \frac{\sigma_{deflection}}{\sigma_E} \right) \\ + \left(\frac{\sigma_S}{\sigma_t} + \frac{\sigma_S}{\sigma_w} + \frac{\sigma_S}{\sigma_P} \right) \end{array} \right\} \tag{8}$$

Based on the procedure summarized in Section 2 and 3, four different ANN metamodels are constructed for computing normal stress, buckling stress, volume and deflection. For this purpose, the training set is provided by 100 sampling using ILHS and testing set is generated using 1000 random sampling on the design variables and uncertain parameters bounds. Metamodeling creation involves making a decision on the appropriate number of layer(s) and the number of neurons in the hidden layer(s) and selecting the best model with minimum MSE. The architecture and approximating capability of these metamodels on training and testing sets are shown in **Table 3**.

As illustrated in **Table 3**, the existence of some areas without enough accuracy is inevitable, so using metamodels in optimization process requires a management strategy. In order to implement the developed management strategy, we need the value of IMSD and FMSD parameters. To make a decision on the value of these parameters, the random testing set is utilized to compute the MSD value of each design point using created metamodels. Increasing in MSD values led to an increase in

Meta-model	Unit	N. neurons in each layer	Train	Test
			MSE	MSE
Normal stress	MPa	[5 5 5]	9.30e−6	81.22
Buckling stress	MPa	[5 5 5]	2.45e−5	2.28e+2
Volume	mm^3	[5 5 5]	5.38e−6	0.0374
Deflection	mm	[5 5 5]	1.22e−5	4.34e+7

Table 3.
Approximation capability of metamodels for two-bar truss problem.

metamodel error. So, during the optimization process, if the amount of the MSD in each individual (i.e. CMSD) be less than PMSD, the metamodel has enough accuracy and can be used to functions evaluation. Contrariwise, when the accuracy of the metamodel is not sufficient (i.e. MSD value is greater than PMSD), the real-model should be used. As stated above, here, the amount of IMSD and FMSD are considered as 2.5 and 0.5, respectively. Also, it is assumed that the adaptive threshold variable k in Eq. (2) be increase from −12.5 to −1 until final optimization generation.

The explained problem is solved through NSGA-II optimizer. The LHS based on the correlation criterion is used to generate the initial population of the optimization process. The optimization setting including population size, generation, crossover, and mutation are 50, 150, 0.8 and 0.15, respectively. The ILHS method with 1000 points is employed for uncertainty propagation and analysis. Finally, the Pareto frontier with two optimality criteria and one robustness criteria using both proposed strategy and real-model simulation is illustrated in **Figure 15**. The number of metamodel/real-model call function and PMSD value are shown in **Figure 16**, iteratively.

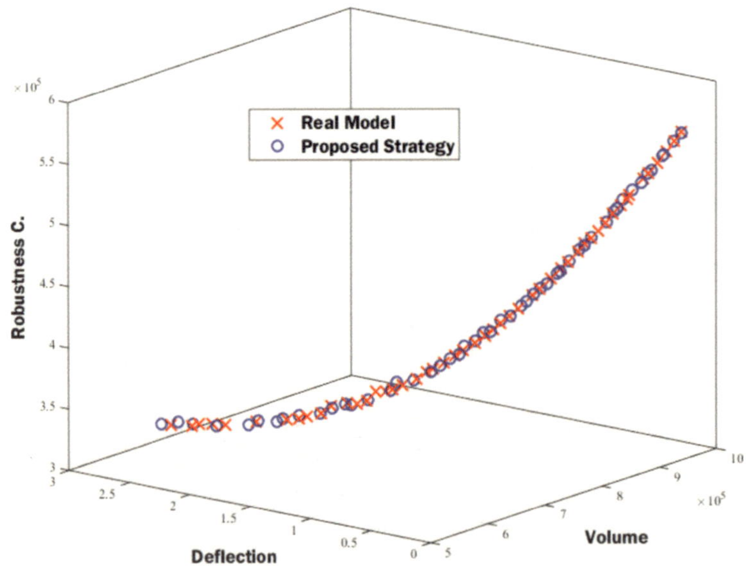

Figure 15.
Pareto frontiers resulted from RDO for two-bar truss problem.

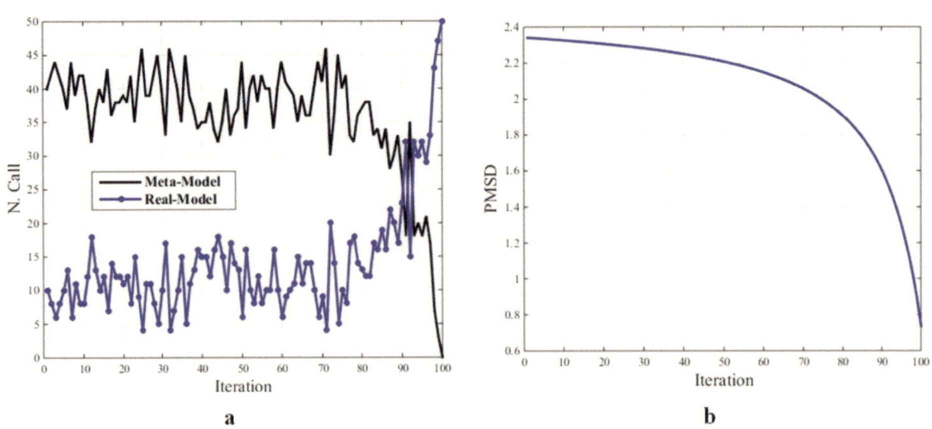

Figure 16.
Number of call functions (a) and PMSD threshold (b) for two-bar truss problem.

According to the iterative results shown in **Figure 16a**, in the early iterations of the optimization process, the number of metamodels call functions are more than real-model ones and in the final iterations, real-model call functions are increased. Therefore, control of call functions in metamodel based optimization process could lead to increasing the accuracy and globality of convergence. In addition, during the optimization process, the tuned adaptive threshold (**Figure 16b**) is slowly decreased to increase the real-model call functions, iteratively.

5. Concluding remarks and future work

Despite recent surrogate-based design books [8, 12, 24] for engineers and extensive investigations conducted in this field, many researchers are still making efforts to push the boundaries of metamodeling. It can be noted that, despite the numerous research carried out over the last few years, the computational complexity still remains as a major challenge of this field of study. Also, today's engineering design problems with multidisciplinary nature are extremely complex (e.g., uncertainty-based multidisciplinary design optimization). Therefore, metamodels management and their accuracy over the design space are another challenges and open fields for research. Therefore, in this chapter, a novel ECS is proposed to improve computational efficiency and make better decisions for function evaluation when facing the metamodel based design optimization problems.

The assessment of the benchmark problems revealed both the efficiency and the effectiveness of the proposed strategy. For all case studies, ANN and Kriging metamodels are used to create metamodels based on ILHS. Also, the ILHS is utilized for uncertainty propagation and analysis. Results illustrate that the proposed strategy could lead to improving the computational efficiency, accuracy, and globality of the convergence process in MBDO problems.

Future researches could be include extensions of the problem to higher dimensional with high fidelity analysis modules and considering different sources of uncertainties. The sensitivity of the proposed strategy to other metamodeling techniques (i.e. RSM, RBF, Kriging, etc.) can be considered in the future. Also, the use of metamodels in co-simulation works to replace high fidelity analysis with inexpensive surrogate models might be an interesting research field. Determining appropriate criteria for extracting or selecting new points to update the metamodels training set during online metamodeling is another challenge of this field of study. Also, the presence of data mining approaches along with computational intelligence methods could provide the basis for the emergence of new metamodeling techniques, which could be very significant.

Author details

Ali Asghar Bataleblu
Department of Aerospace Engineering, K. N. Toosi University of Technology, Tehran, Islamic Republic of Iran

*Address all correspondence to: ali.bataleblu@gmail.com

References

[1] Roshanian J, Bataleblu AA, Ebrahimi M. A novel evolution control strategy for surrogate-assisted design optimization. Structural and Multidisciplinary Optimization. 2018; **58**:1255. DOI: 10.1007/s00158-018-1969-4

[2] Viana FA, Simpson TW, Balabanov V, Toropov V. Special section on multidisciplinary design optimization: Metamodeling in multidisciplinary design optimization: How far have we really come? AIAA Journal. 2014;**52**(4): 670-690. DOI: 10.2514/1.J052375

[3] Jin Y. Surrogate-assisted evolutionary computation: Recent advances and future challenges. Swarm and Evolutionary Computation. 2011;**1**(2): 61-70. DOI: 10.1016/j.swevo.2011.05.001

[4] Sudret B. Meta-models for structural reliability and uncertainty quantification. In: Proc. 5th Asian-Pacific Symp. Stuctural Reliab, Its Appl. Singapore: APSSRA; 2012. pp. 53-76

[5] Simpson TW, Peplinski J, Koch PN, Allen JK. Meta-models for computer-based engineering design: Survey and recommendations. Engineering Computations. 2001;**17**(2):129-150. DOI: 10.1007/PL00007198

[6] Simpson TW, Booker AJ, Ghosh D, Giunta AA, Koch PN, Yang RJ. Approximation methods in multidisciplinary analysis and optimization: A panel discussion. Structural and Multidisciplinary Optimization. 2004;**27**(5):302-313. DOI: 10.1007/s00158-004-0389-9

[7] Wang GG, Shan S. Review of metamodeling techniques in support of engineering design optimization. Journal of Mechanical Design. 2007; **129**(4):370-380. DOI: 10.1115/1.2429697

[8] Forrester AIJ, Keane AJ. Recent advances in surrogate-based optimization. Progress in Aerospace Science. 2009;**45**(1–3):50-79. DOI: 10.1016/j.paerosci.2008.11.001

[9] Booker AJ, Dennis JE Jr, Frank PD, Serafini DB, Torczon V, Trosset MW. A rigorous framework for optimization of expensive functions by surrogates. Structural Optimization. 1999;**17**(1): 1-13. DOI: 10.1007/BF01197708

[10] Tzannetakis N, Van de Peer J. Design optimization through parallel-generated surrogate models, optimization methodologies and the utility of legacy simulation software. Structural and Multidisciplinary Optimization. 2002;**23**(2):170-186. DOI: 10.1007/s00158-002-0175-5

[11] Adams BM, Bohnhoff WJ, Dalbey KR, Eddy JP, Eldred MS, Gay DM, et al. DAKOTA, a Multilevel Parallel Object-Oriented Framework for Design Optimization, Parameter Estimation, Uncertainty Quantification, and Sensitivity Analysis: Version 5.0 User's Manual. Tech. Rep. SAND2010-2183. Sandia National Laboratories; 2009

[12] Tenne Y, Goh CK. Computational intelligence in expensive optimization problems. Vol. 2. Berlin: Springer Science & Business Media; 2010

[13] Horng SC, Lin SY. Evolutionary algorithm assisted by surrogate model in the framework of ordinal optimization and optimal computing budget allocation. Information Sciences. 2013; **233**:214-229. DOI: 10.1016/j.ins.2013.01.024

[14] Sóbester A, Forrester AI, Toal DJ, Tresidder E, Tucker S. Engineering design applications of surrogate-assisted optimization techniques. Optimization

and Engineering. 2014;**15**(1):243-265. DOI: 10.1007/s11081-012-9199-x

[15] Gong W, Zhou A, Cai Z. A multioperator search strategy based on cheap surrogate models for evolutionary optimization. IEEE Transactions on Evolutionary Computation. 2015;**19**(5): 746-758. DOI: 10.1109/TEVC.2015. 2449293

[16] Zhou Q, Shao X, Jiang P, Gao Z, Zhou H, Shu L. An active learning variable-fidelity metamodelling approach based on ensemble of metamodels and objective-oriented sequential sampling. Journal of Engineering Design. 2016;**27**(4–6): 205-231. DOI: 10.1080/09544828. 2015.1135236

[17] Belyaev M, Burnaev E, Kapushev E, Panov M, Prikhodko P, Vetrov D, et al. GTApprox: Surrogate modeling for industrial design. Advances in Engineering Software. 2016;**102**:29-39. DOI: 10.1016/j.advengsoft.2016.09.001

[18] Sun C, Jin Y, Cheng R, Ding J, Zeng J. Surrogate-assisted cooperative swarm optimization of high-dimensional expensive problems. IEEE Transactions on Evolutionary Computation. 2017;**21**: 644-660. DOI: 10.1109/TEVC.2017. 2675628

[19] Sayyafzadeh M. Reducing the computation time of well placement optimisation problems using self-adaptive metamodelling. Journal of Petroleum Science and Engineering. 2017;**151**:143-158. DOI: 10.1016/j. petrol.2016.12.015

[20] Chatterjee T, Chakraborty S, Chowdhury R. A critical review of surrogate assisted robust design optimization. Archives of Computational Methods in Engineering. 2017:1-30. DOI: 10.1007/s11831-017-9240-5

[21] Ryberg AB, Domeij Bäckryd R, Nilsson L. Metamodel-Based Multidisciplinary Design Optimization for Automotive Applications. Linköping: Linköping University Electronic Press; 2012

[22] Zhang SL, Zhu P, Arendt PD, Chen W. Extended objective oriented sequential sampling method for robust design of complex systems against design uncertainty. In: Proceedings of the ASME International Design Engineering Technical Conferences & Computers and Information in Engineering Conference IDETC/CIE. 2012. pp. 12-15

[23] Martinez J, Marti P. Metamodel-based multi-objective robust design optimization of structures. In: 12th International Conference on Optimum Design of Structures and Materials in Engineering; New Forest, UK. 2012

[24] Dellino G, Meloni C. Uncertainty Management in Simulation Optimization of Complex Systems: Algorithms and Applications. New York: Springer; 2015

Section 3

Inspection Techniques

Chapter 3

On-Site Bridge Inspection by 950 keV/3.95 MeV Portable X-Band Linac X-Ray Sources

Mitsuru Uesaka, Yuki Mitsuya, Katsuhiro Dobashi, Joichi Kusano, Eiji Yoshida, Yoshinobu Oshima and Masahiro Ishida

Abstract

Many bridges around the world face aging problems and degradation of structural strength. Visual and hammering sound inspections are under way, but the status of inner reinforced iron rods and prestressed concrete (PC) wires has not yet been confirmed. Establishing a diagnosis method for bridges based on X-ray visualization is required to evaluate the health of bridges accurately and to help with the rationalization of bridge maintenance. We developed 950 keV/3.95 MeV X-band electron linac-based X-ray sources for on-site bridge inspection and visualized the inner structure of a lower floor slab. The information regarding wire conditions by X-ray results was used for the structural analysis of a bridge to evaluate its residual strength and sustainability. For more precise inspection of wire conditions, we applied three-dimensional image reconstruction methods for bridge mock-up samples. Partial-angle computed tomography (CT) and tomosynthesis provided cross-sectional images of the samples at 1 mm resolutions. Image processing techniques such as the curvelet transform were applied to evaluate diameter of PC wires by suppressing noise. Technical guidelines of bridge maintenance using the 950 keV/3.95 MeV X-ray sources are proposed. We plan to offer our technique and guidelines for safer and more reliable maintenance of bridges around the world.

Keywords: bridge inspection, X-ray, nondestructive test, linear accelerator, structural analysis, computed tomography, tomosynthesis, curvelet transform, technical maintenance guidelines

1. Introduction

Maintaining the health of civil infrastructures such as bridges, roads, and tunnels is important to achieve a safe and reliable society [1–3]. Because a large number of concrete structures were built in the age of rapid economic growth in Japan, many of them are approaching their designed life spans. Thus, the development of a reliable health diagnosis technology is currently an urgent task. For example, many of the prestressed concrete (PC) bridges have life spans of 50 years, and they show apparent damage as they approach their designed life spans (**Figure 1**). Approximately 42 and 63% of all bridges will be over 50 years of age by 2021 and 2031, respectively.

(i)

(ii)

(iii)

Figure 1.
Images of degraded bridge and parts. (i) Whole view of the bridge, (ii) degraded surface, and (iii) crack and leak of Ca components.

Normally, concrete bridges are regularly inspected by visual check and hammering tests. If damages or abnormalities are detected, more detailed nondestructive tests (NDTs) must be conducted.

NDT by X-ray radiography is one of the promising technologies for the detailed local inspection of a bridge. X-ray radiography provides high-resolution images of steel wires and rods inside thick concrete bridges. Its penetration capability depends on the thickness of the object and the energy of the X-ray; higher energy X-rays penetrate thicker concrete structures. However, the energy produced by X-ray sources for industrial NDTs is not sufficiently high for the inspection of bridges, because of their thick concrete structure. A low energy source does not provide clear contrast inside the concrete, and in addition, the exposure time required is very long. Some radioisotopes provide high-energy X(γ)-rays. However, they continuously emit X(γ)-rays and operator health becomes another concern. Thus, a new, safer bridge inspection technology using high-energy powerful X-ray sources is needed.

We have developed X-band electron linear accelerator (linac)-based portable X-ray sources with high maximum energies of 950 keV and 3.95 MeV and have been demonstrating X-ray inspection of infrastructures using these sources [4–7]. We have successfully conducted ten on-site inspections using these X-ray sources to date, and we are also working on technology development in laboratories using samples from dismantled bridges.

The main focus of X-ray visualization of a bridge is damage of prestressed concrete (PC) wires and unfilled sheath pipes. Not only the rupture of wires but also the wastage of wires by corrosion reduces the residual strength of a bridge. Therefore, the quantitative evaluation of wire diameters within a resolution of 1 mm is required for X-ray imaging. The grout fill in the sheath pipes of a PC bridge is important to prevent wire corrosion from rainwater and to make the wires and the concrete operate as a composite material. Detecting unfilled sheath pipe areas by X-ray is therefore important for detecting wire damage or for calculating stress imbalances in the bridge. The results of X-ray visualization are used to evaluate the residual strength of a bridge using a reliable numerical calculation method.

The goal of our research is to establish a structural health diagnosis method based on X-ray imaging of the inner structure of a bridge. In this chapter, we present the results of X-ray imaging of an actual bridge still in use and the calculated results of its residual strength through a numerical simulation. Investigation of the validity of X-ray partial-angle computed tomography (CT) or tomosynthesis image reconstruction for precise bridge inspection was also performed.

2. X-ray inspection and structural analysis of a bridge

2.1 950 keV/3.95 MeV X-ray sources

We used X-band (9.3 GHz) linac-based 950 keV/3.95 MeV X-ray sources for the inspection of the actual bridge. The systems are shown in **Figures 2** and **3**, respectively.

In the former, the electrons are accelerated up to 950 keV by radio-frequency (RF) fields. We also adopted the side-coupled standing wave-type accelerating structure. Electrons are injected into a Tungsten target that generates bremsstrahlung X-rays. The generated X-rays are collimated by a Tungsten collimator into the shape of a cone which has an opening angle of 17°. The most important is the X-ray intensity, which is 50 mSv/min at 1 m for a full magnetron RF power of 250 kW. The system consists of a 50-kg X-ray head, 50-kg magnetron box, and stationary electric power source and water chiller unit. The X-ray head and magnetron box are portable, and because they are connected to each other by a flexible waveguide, only the position and angle of the X-ray head are finely tuned. We have optimized the design with respect to X-ray intensity, compactness, and weight. The

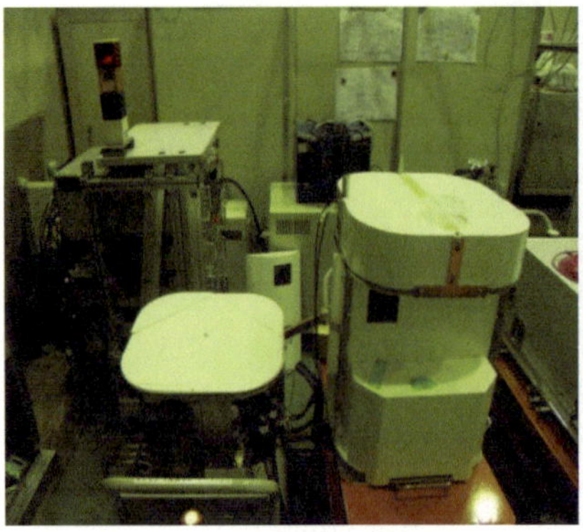

Operating frequency	9300 MHz
Beam energy	950 keV
Beam current	130 mA
Pulse width	2.5 µs
Pulse frequency	330 pulses/s
RF power	250 kW
Electron gun voltage	20 kV
Accelerator length	125 mm
Q-value of the accelerator	7150
X-ray size at target	0.7 mm
X-ray intensity at 1 m	> 50 mGy min^{-1}

Figure 2.
950 keV portable X-band linac-based X-ray source and its specifications. The maximum X-ray energy is 950 keV. The system consists of three units: X-ray head, magnetron, and power units.

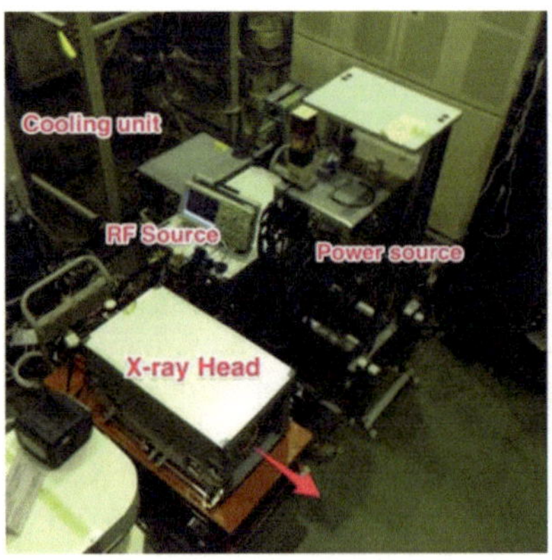

Main unit	Accelerating tube	RF Source	HVPS Control
Weight (kg)	80+62 (Collimator + Accelerating tube)	62	116
Parameters	Electron gun output current 300mA	Frequency 9.3GHz	
	Electron gun voltage 20kV	Pulse width 4μs	
	Beam current 100mA	Repetition rate 200pps	
		RF power output 1.5 MW	

Figure 3.
3.95 MeV portable X-band linac-based X-ray source and its major parameters. The system consists of four units: X-ray head, magnetron, power, and chiller units.

parameters of the 950 keV X-ray source are summarized in the table of **Figure 2**. We place an X-ray detector on the opposite site of the X-ray source between the object and source to detect the transmitted X-rays through the object. We use a flat panel detector (FPD) from PerkinElmer Corporation for the detector.

The 3.95 MeV system is shown in **Figure 3**. This system consists of a 62-kg X-ray head with target collimator of 80 kg, magnetron box of 62 kg, electric power sources of 116 kg, and water cooling system of 30 kg. The X-ray head and magnetron box are portable, and the position and angle of the former are also finely tuned. The X-ray intensity of this system is 2 Gy/min at 1 m.

Calculated attenuations in concrete for the X-rays from the 950 keV/3.95 MeV sources are shown in **Figure 4**. The results indicate that concrete with thicknesses of up to 400 mm and 800 mm can be penetrated by the 950 keV/3.95 MeV sources, respectively.

2.2 Compliance

We comply with Japan's Law Concerning Prevention of Radiation Hazards Due to Radioisotopes and Regulations on Prevention of Ionizing Radiation Hazards when we use the 950 keV/3.95 MeV X-ray sources in the field for on-site bridge inspection. According to the law, an electron beam source below 1 MeV is not

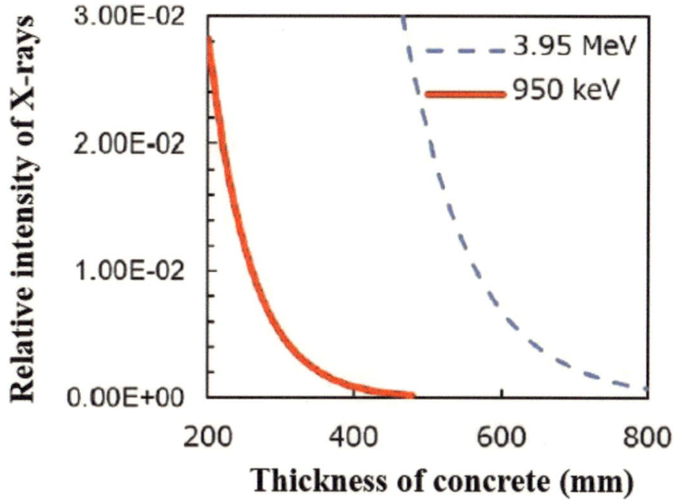

Figure 4.
Calculated results of attenuation for the X-rays in concrete from the 950 keV/3.95 MeV X-ray sources.

an accelerator. Thus, we comply with this regulation. The 950 keV X-ray source is registered with the local agency of labor supervision. We usually operate the source in a radiation-controlled area, which has a radiation safety system complying with the Regulations on Prevention of Ionizing Radiation Hazards. The use of the source outside the controlled area is also allowed. In this case, we temporally set up a controlled area at the measurement site and place sufficient shielding around the source and object to suppress the air dose rate below 1.3 mSv/3 months. Moreover, we have to set a temporal facility boundary of 250 μSv/3 months. Amendment of the law that allows the use of accelerators below 4 MeV only for on-site bridge inspection was implemented in Japan in 2005. After we completed governmental registration as a radiation source, we submitted for permission of use outside the radiation-controlled area. Finally, we performed the on-site inspection under the Regulations on Prevention of Ionizing Radiation Hazards similar to the 950 keV case.

2.3 X-ray transmission imaging

Three types of PC bridges (T-shaped bar type, box-shaped type, and hollow floor bar type) are investigated by the X-ray sources. Possible patterns of X-ray transmission and scanning (partial-angle CT and tomosynthesis) are depicted in **Figure 5**. An X-ray flat panel detector (FPD) and imaging plate (IP) are used for X-ray imaging acquisition. By using the FPD, online measurement in seconds is available so that sparse and fine-tuning of the position of the X-ray sources and detector can be accomplished. Stacking measurement in minutes is more appropriate for the IP. Imaging processing of IPs can be carried out on-site immediately. An aerial work platform and stage are used for measurement of the web and flange parts of a T-shaped bar bridge. The X-ray source can be installed inside a box for bottom floor slab inspection or on a pedestal for upper slabs of box-shaped bar types with the help of a crane. As for hollow floor bar types, the X-ray source is placed on the road or inside a hollow box. Even partial-angle CT and tomosynthesis are applicable for those types of inspections.

Figure 6 shows typical transmission images of the inner structure of the slab of a certain T-shaped bar-type bridge obtained by the 950 keV X-ray source. We successfully observed the inner structure in detail with the linac-based X-ray system.

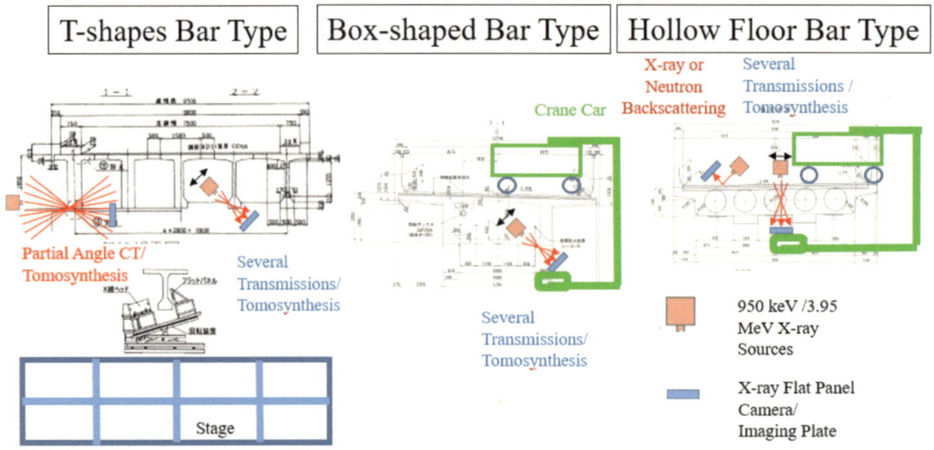

Figure 5.
Typical techniques of X-ray transmission/scanning inspections for bridges of T-shaped bar type, box-shaped bar type, and hollow floor bar type.

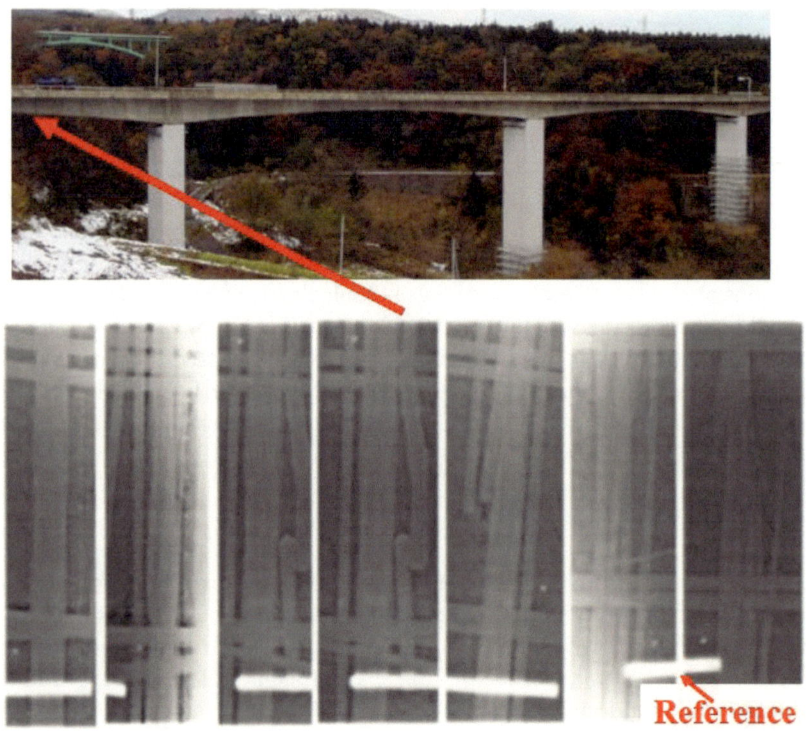

Figure 6.
Series of X-ray images of PC wires in a bottom slab of a box-shaped bar-type bridge acquired by the 950 keV X-ray source. Cutting and thinning of PC wires are observed.

The PC wires were clearly visualized. Cutting and thinning of PC wires are clearly apparent. Reduction of the PC wire cross sections is estimated visually, and images are used for the structural analysis to evaluate any reduction of structural strength quantitatively.

PC wires, sheath, and grout in a web part of other T-shaped bar types obtained by the 950 keV X-ray source are given in **Figure 7**. Even grout filling and missing grout are clearly visible.

Figure 7.
Surface view and X-ray transmission images of near PC wires, sheath, and grout in a web part of a T-shaped bar-type bridge acquired by the 950 keV X-ray source. Cracks and leak of Ca components are visible. Moreover, grout filling and missing grout are clearly observed in the near PC sheath.

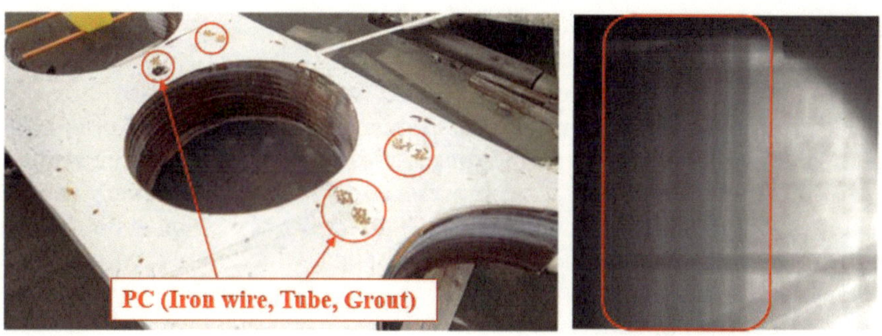

Figure 8.
Cut sample from a hollow flow bar-type bridge and X-ray transmission image acquired by the 3.95 MeV X-ray source. The total thickness of the concrete where the X-rays penetrate is ~400 mm. Several PC wires located deep in concrete are visualized.

Figure 8 shows an X-ray transmission imaged by the 3.95 MeV X-ray source from a cut sample of a hollow flow bar-type PC bridge. Several PC wires located deep in the concrete are visualized.

However, the contrast between the wires and concrete was not sufficiently high. We consider that one reason is noise produced by X-rays scattered by the concrete structure. The probability of Compton scattering of an X-ray becomes higher as the energy of the X-ray becomes higher. If Compton scattering occurs between an X-ray photon and an atom of the material, the photon loses a portion of its energy and changes its direction. Because the original energies of X-rays are high, the energies of scattered X-rays are also high. Thus, many photons may have been scattered within the concrete structure and were detected at the FPD. It is an important task to reduce the image noise caused by scattering X-rays. A possible means to resolve this issue is by introducing a fine pitch metal mesh in front of the FPD to absorb only the X-rays impinging on the FPD and block scattered X-rays coming from other directions. Of course, image processing can be used to reduce noise. This aspect is discussed later.

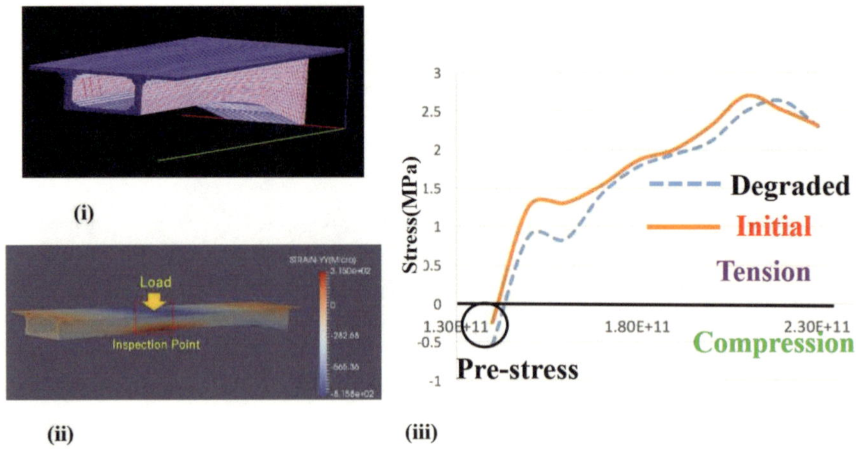

Figure 9.
Typical results of 3D nonlinear structural analysis of iron-reinforced concrete based on DuCOM-COM3 simulation using the cross-sectional reduction based on the measured X-ray transmission images such as those in **Figure 6**. *Mesh model of the finite element method (FEM) and typical load pattern for structural degradation evaluation and stress contour results are shown in (i) and (ii), respectively. Curves of applied moment versus stress for the initial and degraded (measured) states are shown in (iii).*

2.4 Three-dimensional structural analysis

We calculated the residual strength of a block of the bridge using X-ray inspection results and the 3D nonlinear three-dimensional finite element method (FEM) software for reinforced concrete, named DuCOM-COM3 developed by Prof. Hirokazu Maekawa (Department of Civil Engineering, University of Tokyo). The model for FEM analysis is shown in **Figure 9**. We modeled a block of the bridge with an X-ray inspected part at its center, as shown in **Figure 9(i)**. The X-ray inspected part is 1600 mm in the longitudinal direction. The boundary conditions were the moments when the designed load is applied, which were calculated by Newmark's method.

As a result of the FEM analysis, we found that the stress increased about 0.3 MPa at the lower edge of the box girder. This is because the stress resistance was decreased owing to damage of some PC wires. On the other hand, the compressive stress at the upper edge was not affected. The 3D distributions of stress after wire damage are depicted in **Figure 9(ii)**. From the calculations, the load that generates concrete cracks is estimated as 8417 kN for the healthy condition and 8016 kN for the damaged condition. Curves of given moment versus stress for the initial and degraded states are shown in **Figure 9(iii)**. Although the residual strength decreased by approximately 5%, the stress after damage remained within the range of the allowable stress, and the bridge was judged as operable at its current condition.

3. Three-dimensional image reconstruction methods for bridge inspection

3.1 Computed tomography and tomosynthesis for partial angle

Three-dimensional information of the inner structure of the concrete bridge is more helpful than a simple X-ray transmission image if the structure is complicated by densely concentrated wires. In a simple radiography image, the wires are superimposed, and thus a precise evaluation of wire condition is not possible. However, in a three-dimensional image, each wire can be separated, and evaluating the diameters of the wires is easier and more precise.

X-ray CT has been a powerful tool used for this type of purpose in a wide variety of fields, such as medical and industrial applications. In a CT system, the X-ray source and the detector are rotated 360° around the object, and X-ray images are obtained at different angles. Slice images of the object will be reconstructed from the X-ray absorption factors calculated from the images. CT is widely used for the precise inspection of industrial products in the field of NDT; thus, we can expect it is also applicable for bridge inspections using higher energy X-rays.

However, it is usually impossible to widely rotate the source and detector around a target part of a bridge because of limited spaces around the target. We therefore consider applying partial-angle CT for bridge inspections. In partial-angle CT, the source and the detector are rotated less than 360°, and cross-sectional images are reconstructed from those partial-angle projection images. Partial-angle CT formulation is explained as follows, and its coordinate system for the spatial domain is illustrated in **Figure 10(i)**. When the 2D X-ray attenuation constant distribution is $f(x, y)$, the relation between the initial X-ray intensity from the X-ray source, I_0, and input intensity at the detector, I_i, along the arrow in the figure is approximately given as

$$I_0 = I_i \, exp\left\{-\int_s f(x,y)ds\right\}, \tag{1}$$

$$\int_s f(x,y)ds = ln\frac{I_i}{I_0}. \tag{2}$$

Therefore, the Radon integral is measured by

$$p(r,\theta) = \int_{-\infty}^{\infty} f(x,y)ds = \int_{-\infty}^{\infty}\int_{-\infty}^{\infty} f(x,y)\delta(xcos\theta + ysin\theta - r)dxdy. \tag{3}$$

The Fourier transform of $f(x, y)$, $F(\mu, v)$, is given by the measured Radon transform data as

$$\begin{aligned} F(\mu,v) &= \int_{-\infty}^{\infty}\int_{-\infty}^{\infty} f(x,y)\, exp\{-j2\pi(\mu x + vy)\}dxdy \\ &= \int_{-\infty}^{\infty} p(r,\theta)\, exp(-j2\pi\rho r)dr, \end{aligned} \tag{4}$$

where

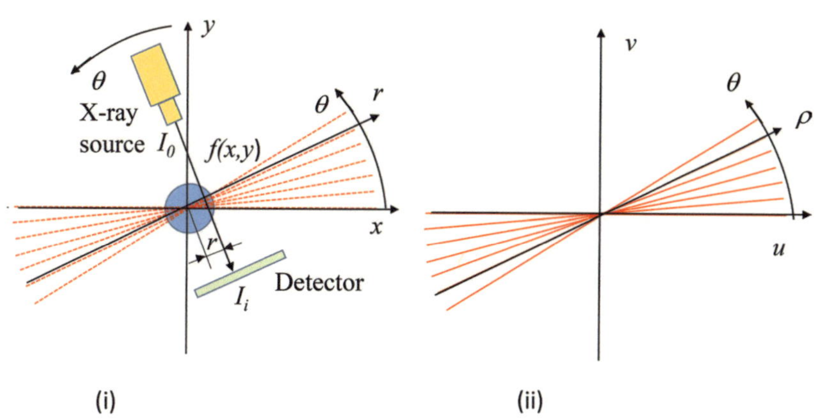

Figure 10.
Coordinate system for the partial-angle CT. (i) Spatial domain, and (ii) Frequency domain.

$$\mu = \rho cos\theta, v = \rho sin\theta,$$

$$\rho = \sqrt{u^2 + v^2},$$

and the coordinate system in the frequency domain is shown in **Figure 10(ii)**. In case of the partial-angle CT, the angle range is rather limited so that the Fourier transform data cannot be obtained in the whole frequency domain. Finally, the original X-ray attenuation distribution is calculated numerical by the filtered back projection method in the following:

$$f(x, y) = \int_{-\infty}^{\infty}\int_{-\infty}^{\infty} F(\mu, v)\, exp\{j2\pi(\mu x + vy)\}d\mu dv. \quad (5)$$

Typical results for limited angles are introduced in Section 3.2.

Another promising technology for obtaining three-dimensional information inside a bridge is tomosynthesis. Tomosynthesis has been used for applications such as breast imaging or dental imaging [8]. In normal tomosynthesis, the source is rotated against the detector with limited angle, and "parallax" X-ray images are obtained. The cross-sectional images of a target object are reconstructed from the parallax images. Tomosynthesis has been used in applications where full-angle CT cannot be applied. Thus, it is also a promising technology for bridge inspections.

As a first step, we experimentally investigated the feasibility of partial-angle CT and tomosynthesis using mock-up samples of bridges.

3.2 Image reconstruction

We have applied the partial-angle CT and tomosynthesis to cut samples from the real PC bridge and acrylic phantoms.

The CT reconstructed results for PC wires in a cut sample of the flange part of a T-shaped bar bridge by scanning at 360, 180, and 90° using the 3.95 MeV X-ray

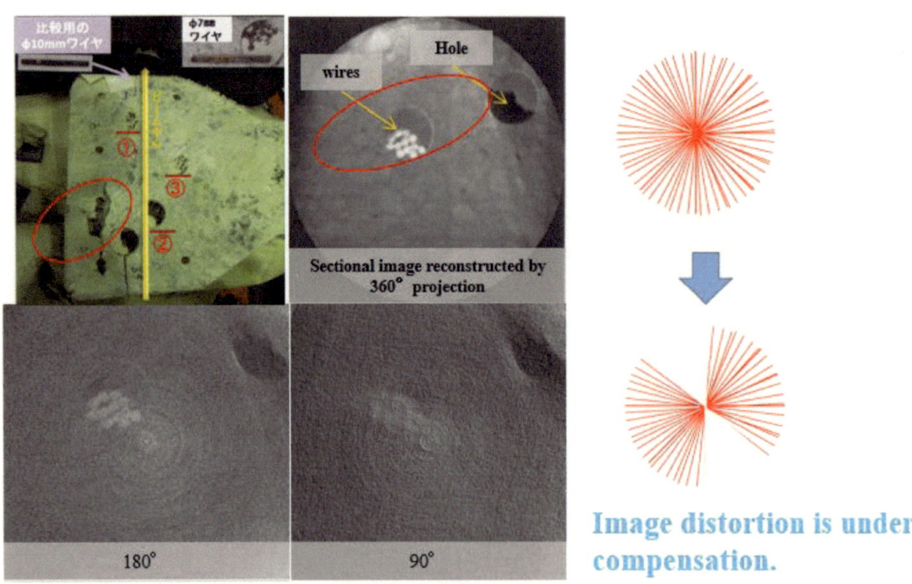

Figure 11.
CT reconstructed results of PC wires in the flange part of a T-shaped bar bridge by scanning at 360, 180 and 90° using the 3.95 MeV X-ray source. All PC wires in a sheath are perfectly reconstructed by 360° scanning.

source are shown in **Figure 11**. All PC wires in a sheath are perfectly reconstructed by 360° scanning. The circular shape is intrinsically deformed to an elliptic shape by partial-angle scanning using the CT algorithm. However, the minor axis diameter of the ellipse is almost the same as the real diameter of the PC wires.

Figure 12 is a reconstructed image of the cross section of an acrylic phantom by tomosynthesis. Although we also observed the deformation of the cross-sectional shapes from circles to ellipses in tomosynthesis, we can estimate the original diameters from the minor axes of the ellipses. We extracted a profile from

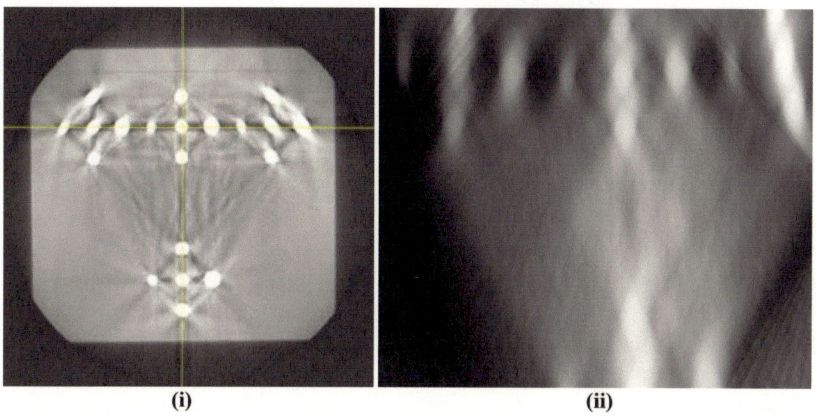

Figure 12.
Cross-sectional X-ray images of the acrylic phantom reconstructed by (i) full-angle CT and (ii) tomosynthesis. The tomosynthesis image was obtained with 25 projections taken in 3° steps.

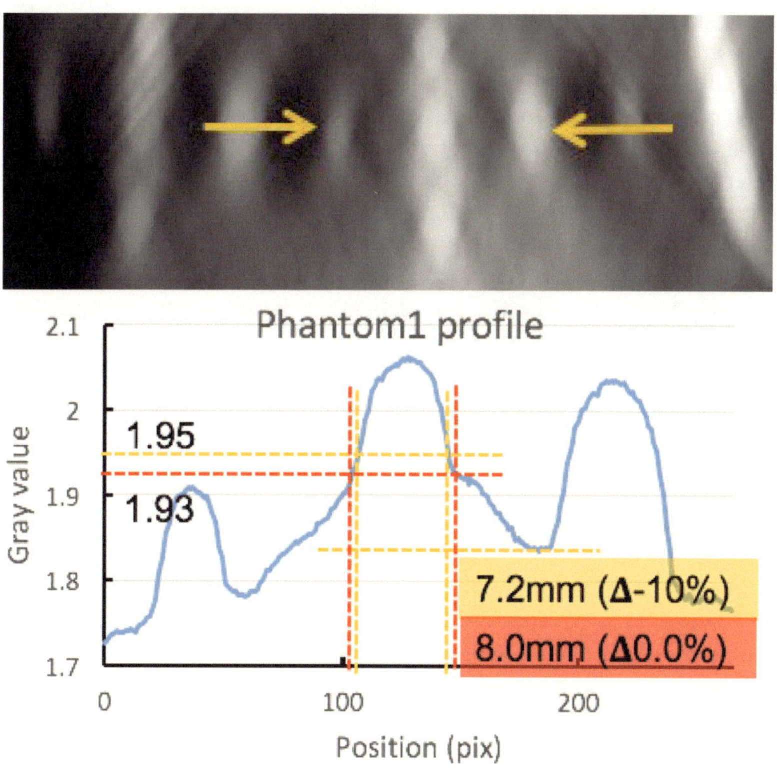

Figure 13.
Gray value plot of a profile in the tomosynthesis image of the acrylic phantom.

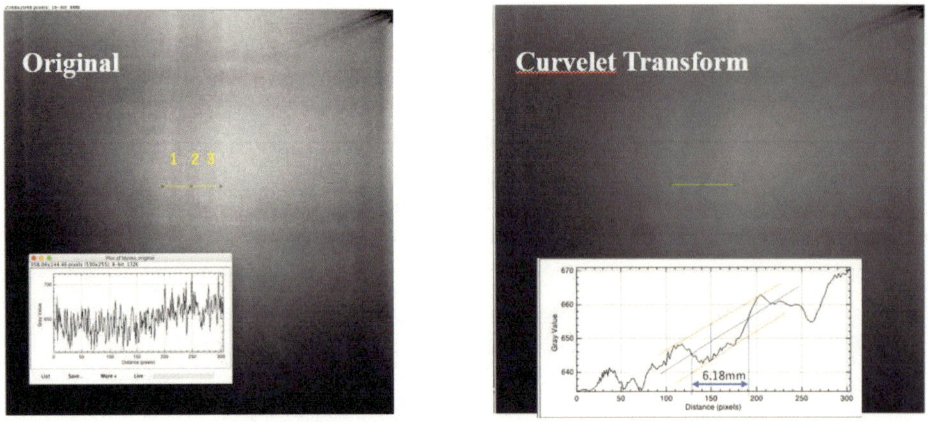

Figure 14.
Image processing using the curvelet transform and evaluation of the diameter of PC wires.

the tomosynthesis image and plotted its gray values (see **Figure 13**). The profile included three rods with different diameters (6, 8, and 10 mm). We were able to estimate the diameter of the center rod (8 mm) within 1 mm accuracy. From the profile plotting, the estimated diameter of the rods was between 7.2 and 8.0 mm.

3.3 Image processing

Software image processing is inevitable for this work. Fourier transform (FT) with low-pass filter, high-pass filter, and band-pass filter is a standard technique. If we apply FT with a low-pass filter, the image becomes blurred, and the boundary between an iron rod and a PC wire is difficult to recognize. As for FT with a high-pass filter, the boundary is emphasized, but the overall view is spotty. When we construct a gray value profile to evaluate the diameter of a rod or PC wire, the profile is noisy. Regarding a band-pass filter, trial and error is necessary to choose an appropriate band. Instead, the wavelet transform is effective for emphasizing local signals. Recently, the curvelet transform has become popular. It is an upgraded wavelet transform fit to emphasize curved and declined boundaries. We applied the curvelet transform to the X-ray transmission images of PC wires of the upper slab of the hollow box bar-type bridge (**Figure 6**). We attempted to evaluate the diameter of one of the PC wires in the somewhat blurred image of the 400 mm slab, which is close to the transmission limit of the X-rays from the 950 keV source. We can observe that spatially high-frequency noises are suppressed and the full width at half maximum (FWHM) can be used to estimate the diameter, which is 6.18 mm for the designed 7 mm wire as shown in **Figure 14**.

4. Highlights from recent inspections of real bridges

4.1 Relationships between states of concrete surface and near PC wires

We check the relationship between the states of a concrete surface and near inner PC wires. We evaluate cut samples from a decommissioned T-shaped bar-type bridge. One example is shown in **Figure 15**. The surface concrete is somewhat degraded and cracked (see **Figure 15(i)**). The cut cross-sectional view and X-ray transmission image of the near PC wires are shown in **Figure 15(ii)** and **(iii)**, respectively. The PC wires look healthy in this case. Because this bridge was located near the sea, the degradation of the concrete was due to salt damage. Furthermore, a load

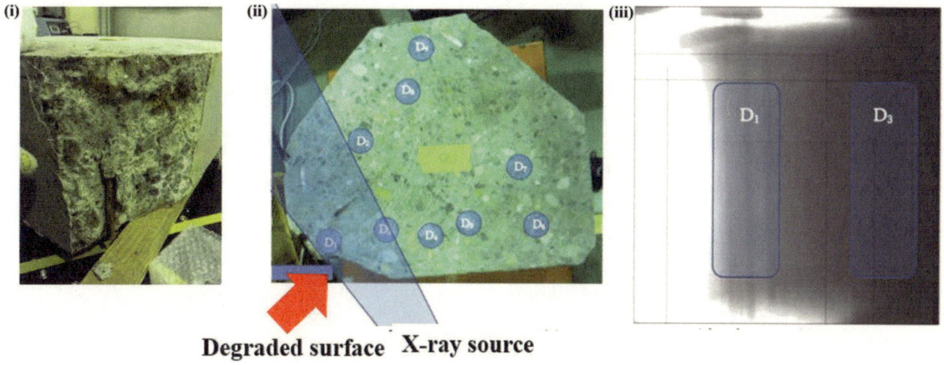

Figure 15.
Typical case of heavily corrupted concrete surface (i) and healthy PC wires in a cut sample from a T-shaped bar bridge. The direction of the X-ray transmission is shown in (ii). X-ray transmission images of PC wires in two sheaths are shown in (iii).

test was performed on this bridge. It was confirmed that its mechanical strength was not degraded, and this mechanically healthy bridge was wastefully decommissioned.

In the above case, inner PC wires were healthy even though the surface concrete was degraded and cracked. Opposite cases involving corroded PC wires in bridges with healthy surface concrete have also been found.

These facts indicate that visible and hammer-sound inspections are not necessarily sufficient for checking the degradation of a bridge's mechanical strength. X-ray inspection is needed to check the state of inner PC wires and to evaluate the mechanical strength of a bridge.

4.2 Filling and missing grout in PC sheath

X-ray transmission images using the 950 keV source in the web part of the T-shaped bar-type bridge in the case of **Figure 7** overlap at the designated location of the two PC sheaths and wires, as shown in **Figure 16**. Grout fills the upper sheath but is missing in the lower. This is the first observation of grout missing after the initial construction. This vacancy may become a puddle of rainwater which would induce corrosion of PC wires in the near future.

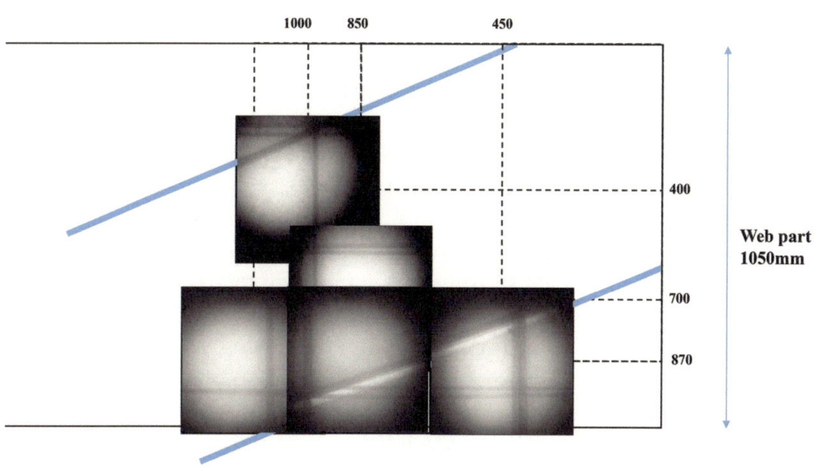

Figure 16.
X-ray transmission images of grout filling and missing grout in the PC sheaths in the web part of a T-shaped bar bridge measured by the 950 keV X-ray source.

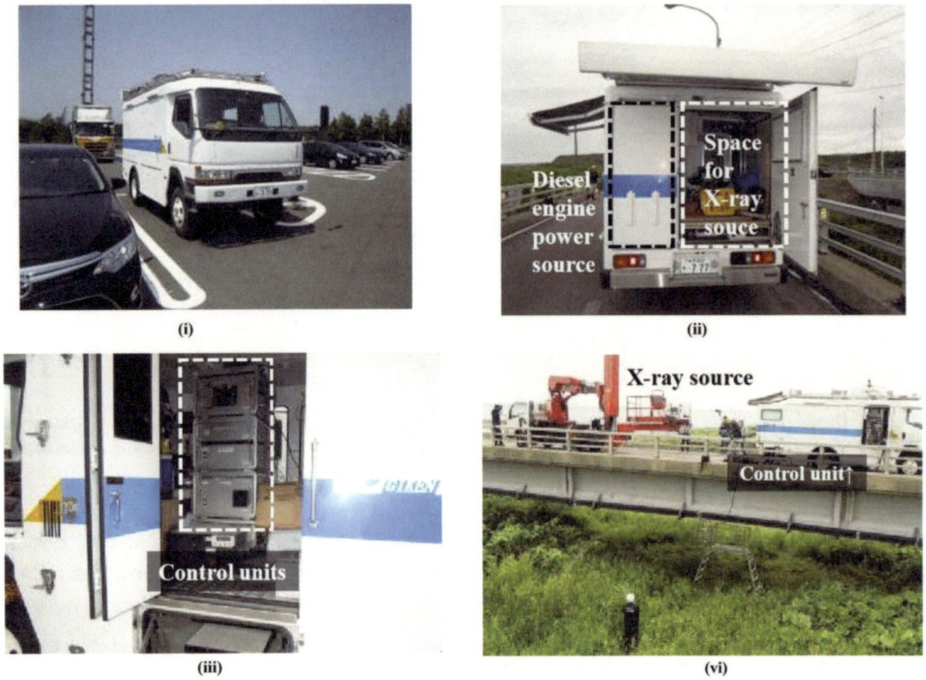

Figure 17.
Special inspection car for the 950 keV/3.95 MeV X-ray sources. (i) Front view, (ii) Back view, (iii) Control units, and (iv) On site set-up.

4.3 Special inspection car

Our collaborator, Kanto Giken Co. (Tokai, Ibaraki, Japan), developed a new special inspection car for the X-ray sources as shown in **Figure 17**. It can carry the X-ray sources, detectors, computers, and inspectors and contains a diesel engine 100 V power source. Some diesel engine power sources have poor quality unstable output voltage and frequency (50 and 60 Hz in east and west Japan, respectively). The electric power source of the magnetron may be sensitive to such instability. To avoid this, we decided to acquire our own reliable diesel engine power source, which we can carry to perform on-site X-ray inspections anywhere in Japan.

4.4 Tuning of position and angle of X-ray sources

The most important procedure of this task are the sparse and fine-tuning of the position and angle of the X-ray sources with respect to the bridge part and the X-ray detectors (FPD and IP) as shown in **Figure 18(i)**, **(ii)**, and **(iii)**. The procedure typically requires approximately 1 h. We use an aerial work platform, stage, and special inspection car for initial settings ((i)), and then by using a rotating function of the X-ray head, we adjust its angle. Finally, fine-tuning among the X-ray source, targeted area, and detector is performed. The whole tuning procedure takes approximately 1 h. This procedure consists of using and setting many devices, including the X-ray sources, detectors, mechanical positioners, and so on.

Usually, we start setting up all devices at 9 am. We then perform sparse and fine-tuning of the position and angle of the X-ray sources and detectors and begin taking real measurement at 11 am. We scan several parts until 3 pm. Finally, dismantling and storing the equipment occurs between 3 pm and 5 pm. Targeted bridge parts are

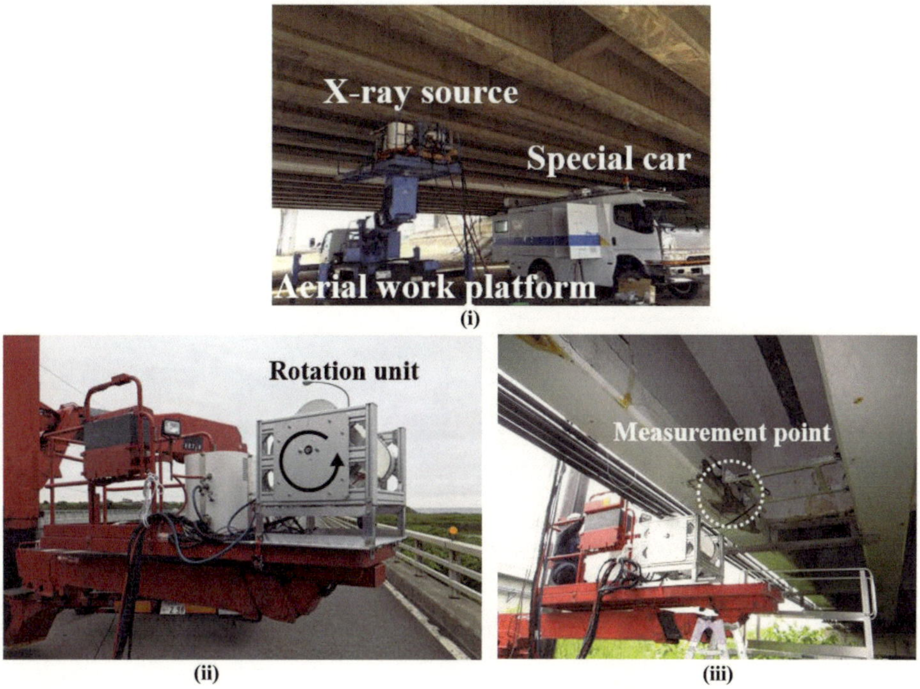

Figure 18.
Initial setup (i), control of angle (ii), and fine positioning (iii) of the 950 keV X-ray source with an aerial work platform.

completely different depending on the three types of bridges (T-shaped bar, box-shaped bar, or hollow floor bar). Thus, we are always improving and upgrading not only the X-ray sources but also our devices and software to better deal with many complicated situations.

Damage to bridges in northern and highland areas in Japan is serious because water in the concrete of a bridge freezes and expands, causing cracks in the concrete. It may snow in those areas in winter. Therefore, inspections should not be carried out during winter. It often rains from June through October in the semitropical climate areas of Japan. We have to be prepared for rain and high humidity, so we use waterproof housing for electric sources and perform very careful equipment treatments and inspections to avoid electric breakdown.

5. Guidelines for special inspections using 950 keV/3.95 MeV X-ray sources

The Public Works Research Institute and the University of Tokyo are developing new technical guidelines for special inspections of bridges using 950 keV/3.95 MeV X-ray sources. An overview is provided in **Figure 19**. First, visual and hammer-sound inspection screening should be performed based on regular inspection guidelines. Advanced hardware and software techniques such as drawn and acoustic analysis are adopted in this step. If degraded parts are found, the special X-ray transmission inspection is performed using the 950 keV or 3.95 MeV X-ray sources, depending on the thickness of the concrete containing the degraded parts. Here, the states of PC wires and rods as affected by corrosion, cuts, and reduction of cross sections are quantitatively evaluated with spatial resolution of 1 mm. Then, 3D nonlinear structural analysis is performed to evaluate

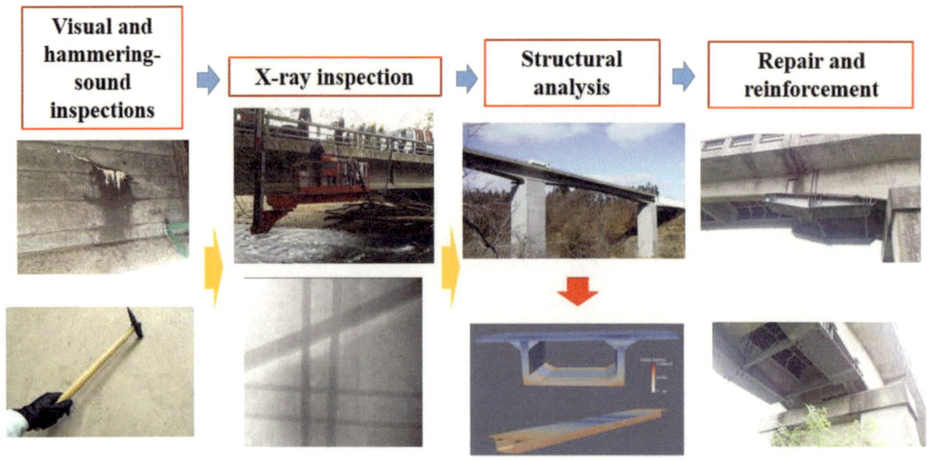

Figure 19.
Guidelines for special X-ray transmission inspection using 950 keV/3.95 MeV X-ray sources accompanied with visual and hammer-sound inspections, structural analysis, final repair, and/or reinforcement.

the degradation of the structural strength quantitatively. Based on this evaluation, repair, reinforcement, or other decisions should be reviewed. Several inspection industries are joining our project and technical transfer is being promoted. We hope to soon apply these guidelines to all aged bridges in Japan and finally across the world.

6. Summary

We have been developing a new X-ray diagnostic method for social and industrial infrastructures using linear accelerator-based X-ray sources. Our 950 keV/3.95 MeV X-ray sources have been applied to many different cases involving on-site X-ray inspection of bridges in Japan. We are currently undertaking on-site inspections of actual bridges. We have demonstrated X-ray inspection of an actual bridge still in use. Clear X-ray transmission images inside the concrete were successfully obtained in the demonstration. The information regarding PC wire conditions from X-ray images was applied to the structural analysis using the finite element method to evaluate the residual strength of a bridge. We found that the residual strength of the bridge in question had decreased by approximately 5% from its original state based on application of the designed load.

We also studied three-dimensional image reconstruction methods that can be applied to bridge inspections. Because of the limitations in spaces and rotation angles involved in actual on-site inspections, full-angle CT is normally not possible to apply for bridge inspections. Thus, we investigated the effectiveness of partial-angle CT and tomosynthesis for bridge inspections. Although the cross-sectional shape of a wire or rod was deformed from its original circular shape to an ellipse-like shape, we could estimate the diameter of the rod or wire from the length of the minor axis of the ellipse. The estimated diameter of a steel rod in a tomosynthesis image was in good agreement with its real value.

Further studies are required to realize the practical use of this X-ray bridge inspection method. We should continue the demonstrations involving on-site bridge inspection for different types of bridges and evaluate the effectiveness of the special X-ray transmission inspection using the 950 keV/3.95 MeV X-ray sources based on the new technical guidelines.

Acknowledgements

This work was supported by the "Infrastructure Maintenance, Renovation, and Management program" of Cross-ministerial Strategic Innovation Promotion Program (SIP), Cabinet Office, Government of Japan. The authors thank for Mr. Kentaro Murata of XIT Co. for the partial CT analysis.

Author details

Mitsuru Uesaka[1]*, Yuki Mitsuya[1], Katsuhiro Dobashi[1], Joichi Kusano[2], Eiji Yoshida[3], Yoshinobu Oshima[3] and Masahiro Ishida[3]

1 Nuclear Professional School, School of Engineering, The University of Tokyo, Ibaraki, Japan

2 Accuthera Inc., Kawasaki, Kanagawa, Japan

3 Center for Advanced Engineering Structural Assessment and Research, Public Works Research Institute, Ibaraki, Japan

*Address all correspondence to: uesaka@tokai.t.u-tokyo.ac.jp

References

[1] Chang P et al. Review paper: Health monitoring of civil infrastructure. Structural Health Monitoring. 2003;**2**(3):257-267

[2] Cao H et al. Form-finding analysis of suspension bridges using an explicit iterative approach. Structural Engineering and Mechanics. 2017;**62**(1):85-95

[3] Qin S et al. Dynamic model updating for bridge structures using the kriging model and PSO algorithm ensemble with higher vibration modes. Sensors. 2018;**18**(6):1879

[4] Uesaka M et al. 950 keV, 3.95 MeV and 6 MeV X-band linacs for nondestructive evaluation and medicine. Nuclear Instruments and Methods in Physics Research A. 2011;**657**(1):82-87

[5] Natsui T et al. Development of a portable 950 KeV X-band Linac for NDT. American Institute of Physics Conference Proceedings Series. 2009;**1099**:75-78

[6] Uesaka M et al. Commissioning of portable 950 keV/3.95 MeV X-band linac X-ray source for on-site transmission testing. E-Journal of Advanced Maintenance. 2013;**5**(2):93-100

[7] Uesaka M et al. On-site non-destructive inspection of the actual bridge using the 950 keV X-band electron linac X-ray source. Journal of Disaster Research. 2017;**12**(3):578-584

[8] Niklason LT et al. Digital tomosynthesis in breast imaging. Radiology. 1997;**205**:399-406

Section 4

Fatigue Assessment

Chapter 4

Development of a Fatigue Life Assessment Model for Pairing Fatigue Damage Prognoses with Bridge Management Systems

Timothy Saad, Chung C. Fu, Gengwen Zhao and Chaoran Xu

Abstract

Fatigue damage is one of the primary safety concerns for steel bridges reaching the end of their design life. Currently, US federal requirements mandate regular inspection of steel bridges for fatigue cracks; however, these inspections rely on visual inspection, which is subjective to the inspector's physically inherent limitations. Structural health monitoring (SHM) can be implemented on bridges to collect data between inspection intervals and gather supplementary information on the bridges' response to loads. Combining SHM with finite element analyses, this paper integrates two analysis methods to assess fatigue damage in the crack initiation and crack propagation periods of fatigue life. The crack initiation period is evaluated using S-N curves, a process that is currently used by the FHWA and AASHTO to assess fatigue damage. The crack propagation period is evaluated with linear elastic fracture mechanic-based finite element models, which have been widely used to predict steady-state crack growth behavior. Ultimately, the presented approach will determine the fatigue damage prognoses of steel bridge elements and damage prognoses are integrated with current condition state classifications used in bridge management systems. A case study is presented to demonstrate how this approach can be used to assess fatigue damage on an existing steel bridge.

Keywords: fatigue, fatigue damage, structural health monitoring, damage prognoses, fatigue assessment, bridge management systems, condition ratings

1. Introduction

In 2013, the American Society for Civil Engineers (ASCE) released an updated Infrastructure Report Card that found nearly 25% of the nation's bridges to be either structurally deficient or functionally obsolete. A bridge is considered structurally deficient (SD) when it is in need of significant maintenance, rehabilitation, or replacement due to deteriorated physical conditions and is considered functionally obsolete (FO) when it does not meet current standards, such as vertical clearances or lane widths. To make these condition assessments, the Federal Highway

Administration uses information from inspection reports that are hosted by state and federal bridge management systems (BMS). BMS are heavily dependent on field inspectors, who collect information on bridge elements and bridge components, evaluate their condition, and enter this data into the BMS database. Among the various tasks of BMS, field inspection is the most essential in evaluating the current condition of steel bridges, which are vulnerable to fatigue-induced damage: the process of material degradation and/or cracking by repeated loads. Fatigue damage occurs over a long period of time and is the primary failure mechanism in steel bridges reaching their original design life [1]. Fatigue damage is largely dependent on the size of the traffic loadings, the frequency of the loads, and the type of detail under examination [2]. The damage usually initiates at the fatigue-prone areas of the bridge: the bridge connections, attachments, and details, such as welds connecting connection plates to steel girders. The defects begin to grow under repetitive loads until a bridge inspector finds the crack in a visual inspection. If the crack is not attended to, it will continue to grow until the structural component is capable of fracture and is also considered to be at the end of its total fatigue life.

Currently, the Federal Highway Administration (FHWA) uses fatigue life estimations to predict the performance of steel bridge members [3]. These fatigue estimates describe the onset of a crack by correlating the magnitude of the stress ranges with the number of load cycles the member has experienced. However, once cracking has occurred, there are no federal or state specifications for crack analysis or crack growth predictions. The fatigue life assessment can be more accurately characterized when crack growth analysis is also included in the assessment. This paper presents a fatigue life assessment method that combines the stress-cycle approach, currently used in AASHTO LRFD Bridge Design Specifications 2014, with a fracture mechanics approach. The damage accumulation results are integrated with current condition state classifications used in BMS.

2. Fatigue life assessment modeling

The fatigue life of a member is the number of load cycles a member can endure before confronting the structure's serviceability limit state. Within a structure's fatigue life, a structure is considered to experience deterioration in two different periods in time: crack initiation and crack propagation. The crack initiation period describes the time when cracks are just beginning to initiate from points of stress concentrations in structural details. Starting with an inclusion in the material, an initial microscopic crack grows a microscopically small amount in size each time a load is applied. The crack initiation period ends when a microscopic crack reaches a predefined critical crack size, typically a crack that is visible in size. The initiation period covers a significant part of the fatigue life. Once a fatigue crack has initiated, applied repeated stresses cause propagation, or growth, of a crack across the section of the member until the member is capable of fracture. The crack propagation period ends when a crack has reached a critical size or final crack size, determined from the material fracture toughness. When a structure has experienced a crack size at the end of the propagation life, the structure is capable of fracture and is also considered to be at the end of its total fatigue life. It is technically significant to consider the crack initiation and crack propagation stages separately because the practical conditions that have a large influence on the crack initiation period are different from the conditions that will influence the crack propagation period [4].

2.1 Fatigue crack initiation period

The crack initiation period corresponds to the onset of a fatigue crack in a component under traffic loads due to an applied stress. To properly account for the dynamic effects in traffic loads, it is necessary to gather a realistic set of data on the stress history that depends upon bridge traffic [5]. This can be accomplished through structural health monitoring (SHM).

2.1.1 Structural evaluation using structural health monitoring

A SHM system gathers real-time measurements of a structure behavior under the effects of varying vehicle weights and their random combinations in multiple lanes. Therefore, the measured strain data reflects the loading conditions in the particular location of the strain gage. SHM methodologies can be divided into two main categories: a statistical/data model-based approach and a physical model-based approach. In the statistical model-based approach, only the measured response of the structure is considered for an assessment, while a physical model-based approach concentrates on the understanding of the structure from its physical model, and a finite element analysis is frequently employed and validated through SHM [6]. In the physical model-based approach, the field measurements verify and validate the finite element models, and a simulation of traffic loads can be used to conduct a structural damage assessment.

To accurately characterize load histories, the content of a measured signal should be summarized and quantified in a meaningful way. The rainflow cycle counting method is recognized as the most accurate way of representing variable amplitude loading [7] and is preferred for statistical analysis of load-time histories, as described in the standard of the American Society for Testing and Materials [8]. Rainflow counting method is advantageous to other range counting methods because it offers realistic counting results while preserving the amplitudes of the acquired stress ranges. As part of the cycle counting process, it is customary to remove small oscillations that are negligible contributors to fatigue damage. Further, the stress ranges caused from smaller vehicles are often considered negligible compared to trucks. This is not only established in *AASHTO Standard Specifications for Highway Bridges* [9], but the NCHRP Report, *Fatigue Evaluation of Steel Bridges* [10], also pays distinct attention to truckloads when estimating fatigue life, stating "the effective stress range shall be estimated as either the measured stress range or a calculated stress range value determined by using a fatigue truck as specified in the AASHTO LRFD Bridge Design Specification 2014 [11]." Because of the significance of truckloads compared with smaller vehicular passages, it is rational to neglect stress cycles below 1 ksi [12].

2.1.2 Bridge global model

Alongside structural health monitoring, a three-dimensional finite element global model can be developed for linear elastic structural analyses. For a typical steel highway bridge, the global model includes the deck, girders, connection plates, and the cross frames to the girders. The global model contains only the main components of the bridge and is primarily used for modal analysis, finding the displacement output of the whole bridge, and critical fatigue location determination known as hotspots, i.e., the locations of known high tensile strength. Field measurements were taken to calibrate the finite element model, accelerometers were used to capture the bridge frequency, laser sensors and potentiometers were used to measure the dynamic deflection of the bridge, and strain gages were used on connection plates to capture the stresses of bridge components. The simulation of truckloads on the global model will output the stresses of all the components on the bridge.

2.1.3 Global simulation modeling

Global simulation modeling uses a three-dimensional model of a bridge with a traffic simulation to estimate fatigue damage. The fidelity of the fatigue assessment is dependent on the accuracy of the traffic load model and the accuracy of the structural model. Since larger loads (i.e., truckloads) are major contributors to fatigue damage and the global simulation model requires computational complexity, the traffic simulation only considers truck loading data for the fatigue assessment. There are two main components of truckloads to consider: the loading configuration (i.e., axle weights and axle spacing) and the traffic patterns. Weigh stations and traffic monitoring systems are often used by State Transportation Departments to acquire loading configuration and traffic pattern data. This data can be used to develop a traffic load simulation, also referred to as the truckload spectra.

To generate load configuration data, the *Guide Specifications for Fatigue Evaluation of Existing Steel Bridges* [13] recommends collecting data through weigh station measurements. A weigh station is a checkpoint equipped with truck scales. Trucks and commercial vehicles are subject to passing the scales at a very low speed and return to the highway after inspection. Data collected from weigh station measurements includes the number of axles and the axle spacing. The collection of truck traffic data at weigh stations can be used to calculate the effective gross weight of the truck spectra:

$$W = \left(\Sigma f_i W_i^3\right)^{1/3} \tag{1}$$

where f_i is the fraction of gross weights within an interval and W_i is the midwidth of the interval.

The traffic patterns are another influence to fatigue damage. The actual traffic flow through a bridge is affected by the traffic on the connecting roadways. Automatic traffic recorders can be used to realistically capture the actual traffic patterns, such as vehicle speed, lane distribution, and vehicle position. Time-varying vehicular count data combined with weigh station measurements are used to develop a probabilistic-based truck simulation model. After obtaining the time-history spectra, the fatigue life and the remaining fatigue life for this detail can be calculated as a function of stress range and number of cycles. Detailed traffic load simulation is reported in a separate companion paper, *Fatigue Assessment of Highway Bridges under Traffic Loading Using Microscopic Traffic Simulation*.

2.1.4 Crack initiation life prediction

The crack initiation period is characterized by the S-N curve. S-N curves are used to relate the stress range (S) vs. number of loading cycles (n_i) and ultimately define the fatigue life of the material. S-N curves comprise the influence of material, the geometry of the local structure, and the surface condition. Failure for the crack initiation period is defined by a crack that is of a critical size. Until the onset of this fatigue crack, the specimen can be characterized by the amount of current fatigue damage in terms of its fatigue life. So, the specimen may be at x% of its fatigue life, or the specimen can be classified to have (100 – x)% remaining useful life. This damage may not be visible upon inspection but is still present in the material.

Since the data in S-N curves were developed under constant amplitude cyclic loading, an effective stress range should be calculated to equivalently represent the variable amplitude cyclic loading on bridge structures. The effective stress range for a variable amplitude spectrum is defined as the constant amplitude stress range that would result in the same fatigue life as the variable amplitude spectrum. For steel structures, the root mean cube stress range (Eq. 2) is calculated from a variable amplitude stress range histogram and is used with the constant amplitude S-N curves for fatigue life analyses [14]:

$$S_{re} = \left(\Sigma \gamma_i S_{ri}^3\right)^{1/3} \tag{2}$$

where S_{ri} is the midwidth of the *i*th bar, or interval, in the frequency-of-occurrence histogram, 3 is the reciprocal of the slope in the constant S-N curve, and γ_i is the fraction of stress ranges in that same interval [15].

2.1.5 Damage accumulation: crack initiation period

The damage accumulation of crack initiation period, d_i, is calculated by comparing the effective stress range to the predefined laboratory values of specimens which are used to construct the S-N curve. Thus, the cumulative damage from the crack initiation life is written as a percentage of the fatigue life by dividing the number of current cycles at the effective stress range, N_e, by the number of stress cycles to fatigue failure, N_f:

$$d_i = {}^{N_e}\!/_{N_f}(100)\,\%\text{damage} \tag{3}$$

2.2 Fatigue crack propagation period

In the crack propagation period, the crack is considered to be a macro-crack and is now growing through the material. The rate of this crack growth is highly dependent on the material type. While the nature of the material cracking is a nonelastic deformation, the region beyond the crack (at the crack tip) experiences a linear elastic stress field under load.

2.2.1 Linear elastic fracture mechanics

Because the stresses at the crack tip are so small in fatigue problems, the plastic zone is limited, and linear elastic fracture mechanics (LEFM) can be used to assess fatigue crack propagation. Paris model is most widely used model in linear elastic fracture mechanics for the prediction of crack growth. In this model, the range of the stress intensity factor is the main factor driving the crack growth with two parameters C and m that reflect the material properties:

$$\frac{da}{dN} = C\,(\Delta K)^m \tag{4}$$

where a is the initial crack size, N is the number of fatigue loading cycles, C and m are material properties, and ΔK is the stress intensification factor. For a given initial crack size, once the crack growth rate is determined, then the existing crack size can be easily calculated through a summation over crack size increments starting from the known size.

2.2.2 Stress intensity factor

The stress in the local crack tip is described as a function of the applied stress in the form of a stress intensity factor (SIF). SIFs are used to describe the severity of a stress distribution around a crack tip, the rate of crack growth, and the onset of fracture [16]. Even at relatively low loads, there will be a high concentration of stress at the crack tip, and plastic deformation can occur [17]. The simplest form to describe the "intensity" of a stress distribution around a crack tip can be written as

$$K = \beta S \sqrt{\pi a} \tag{5}$$

where S is the remote loading stress, a is the crack length, and β is a dimensionless factor depending on the geometry of the specimen or structural component. One important feature this equation illustrates is that the stress distribution around the crack tip can be described as a linear function.

For many ordinary cases of cracking, the calculations of stress intensification factors for various crack geometries and loading cases have already been computed and can be obtained from previously published literature, e.g., elliptical cracks embedded in very large bodies [4]. However, for cases with more complex geometries, more accurate K values should be independently calculated. Finite element modeling (FEM) offers a variety of techniques and efficient computation and has proven to offer satisfactory results for the stress intensification factors [4]. In finite element models, the crack is treated as an integral part of the structure and can be modeled in as much detail as necessary to accurately reflect the structural load paths, both near and far from the crack tip.

2.2.3 Fracture toughness

When the crack grows to a particular size, the stresses at the crack tip are too high for the material to endure, and fracture takes place. This critical stress intensity value is more often referred to as the fracture toughness, K_{Ic}, where I denotes opening mode and c represents critical. Fracture toughness is a measured material property, just like Poisson's ratio or Young's modulus, and is usually measured through standard compact specimens. The fracture toughness is used to describe the ability of an already cracked material to resist fracture or to indicate the sensitivity of the material and the material's susceptibility to experiencing cracks under loading [4]. Thus, SIFs can be compared with the fracture toughness variables to determine if the crack will propagate and to determine the size of crack a material can endure until fracture [18]. When the applied stress intensity equals or exceeds the material fracture resistance, K_{IC}, fracture is predicted.

2.2.4 Crack propagation period cumulative damage

Models that predict fatigue crack growth propagation emphasize that crack growth is largely dependent on the cycle-by-cycle process. Prediction models are referred to as interaction models and non-interaction models. Interaction effects imply that the crack growth rate in a particular cycle is also dependent on the load history of the preceding cycles rather than an independent effect from one cycle. A non-interaction prediction model is used if the interaction effects in the variable amplitude history are assumed to be absent. In a non-interaction model, crack growth in each cycle is assumed to be dependent on the severity of the current cycle only and not on the load history in the preceding cycles. While it is expected that a non-interaction model will lead to a more conservative life prediction than models that account for interaction effects, considering interaction effects account for retardation in crack growth, a non-interaction model can provide quick and useful information about fatigue crack growth behavior, particularly crack growth rates [4]. The non-interaction prediction model leads to a simple numerical summation in Eq. (6), where $\Delta a = da/dN$:

$$a_n = a_0 + \sum_{i=1}^{i=n} \Delta a_i \tag{6}$$

The accumulation of damage for fatigue crack growth models is consequent of the change in crack size, a, where a_0 is the initial crack size, Δa_i is the change in crack size per cycle, and a_n is the updated crack size [4]. Thus, the cumulative damage from the fatigue crack propagation period, d_p, is written as a percentage of the fatigue life by dividing the current crack size, a_n, by the critical crack size at failure, a_{crit}:

$$d_p = {}^{a_n}/_{a_{crit}}(100)\ \%\ \text{damage} \tag{7}$$

3. Damage prognoses fatigue life

The assessment for the crack initiation period and the assessment for the crack propagation period can be combined to determine a damage prognosis, D_{Total}, for the entire fatigue life:

$$D_{Total} = \begin{cases} \alpha_I d_i, N_e \le N_f \\ \alpha_I d_i + \alpha_P d_p, N_e > N_f \end{cases} \tag{8}$$

where N_e is the number of cycles the element has currently experienced, N_f is the number of cycles to failure, d_i is obtained from Eq. (3) and d_p is obtained from

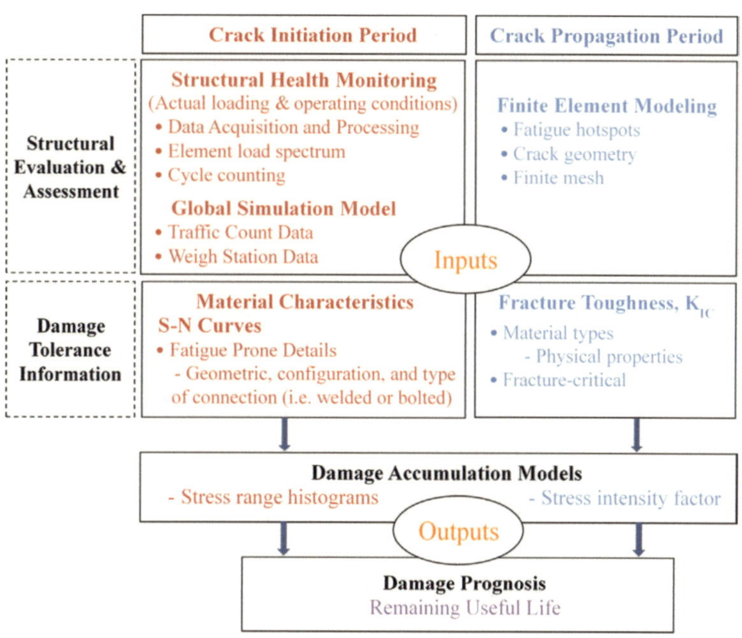

Figure 1.
Fatigue damage prognoses with structural health monitoring.

Eq. (7), and α_I and α_P are rate adjustment factors since the crack initiation period and crack propagation period are not equal in time. These factors can be altered to reflect the rate of damage. **Figure 1** displays the various aspects of fatigue analyses that are considered in the derivation of a fatigue damage prognosis. The diagram summarizes the analyses that are detailed in the preceding sections of this paper. As seen in the diagram, the damage accumulation model that defines crack initiation is informed by a structural evaluation, which can be conducted by means of a global simulation model that is validated with structural health monitoring. SHM gathers information about the actual load distributions and operating conditions of the bridge components. This information is processed and evaluated with damage tolerance information, which describes the material characteristics and material properties, such as the number of stress cycles a structural element can endure before cracking. The damage accumulation model that defines the crack propagation period is informed by finite element models of fatigue hotspots with existing cracks. The finite element modeling provides insight of the stress rate at the crack tip. Fracture toughness is then used to determine the critical crack size, at which the structure is described to be at the end of its fatigue life. Ultimately, the damage accumulation models in the crack initiation period and crack propagation period are used to determine the structure's damage prognosis (remaining useful life).

3.1 Integration of damage prognosis with bridge management systems

Currently, most US state Departments of Transportation (DOTs) report their bridge inspection findings using AASHTO Pontis software, which poses the guidelines for capturing damage of bridge elements. The conditions of bridge elements are categorized into element condition states to reflect these damages. The AASHTO Pontis software is most useful for state DOTs, since it provides an internal tool for mapping the element condition states back into the national condition ratings. **Table 1** summarizes the four condition states related to fatigue damage. These condition states are found in the *Maryland Pontis Element Data Collection Manual* [19].

The condition states in **Table 1** can be used with the fatigue life curve (**Figure 2**) to gather quantitative information of the fatigue life. An element in condition state one is considered a new element or in "like new" condition; it has no fatigue damage present. This element falls within the early stages of the crack life-initiation period. Condition state two recognizes fatigue damage. This damage could be found from a stress-cycle analysis that showed the structure was nearing the end of the crack initiation life or could be the result of a visual inspection from of a small crack that is not considered to be in immediate need of repair. An element in condition state two will be approaching the critical crack size of the crack initiation period and is merging into the crack propagation period. Thus, an element is in the propagation period in condition state three, which explicitly calls for additional analyses. In many state DOTs, it is suggested that deterioration modeling be used to assess the fatigue damage and evaluate the probability of transitioning from condition states [19]. A stress-cycle history can be used to obtain information about the daily or yearly cycle count and stress ranges on the structure. In the event there is enough information about the crack, crack growth models can be used to obtain information about the crack growth rate. This is particularly important information to obtain if the fatigue damage is on a primary component of the structure. Finally, an element in condition state four is in need of immediate rehabilitation or replacement. Analysis can still be used to understand the problem with this section of the bridge to make appropriate changes and to increase the bridge life.

A description of the national bridge element condition states is described in **Table 1** and is used in parallel with the FHWA Bridge Preservation Guide, which

National bridge element condition states				
Defect	**Condition state 1 (good)**	**Condition state 2 (fair)**	**Condition state 3 (poor)**	**Condition state 4 (severe)**
Cracking/fatigue	None	Fatigue damage	Fatigue damage (Analysis warranted)	Severe fatigue damage
		Fatigue damage exists but has been repaired or arrested. The element may still be fatigue prone	Fatigue damage exists which is not arrested. Condition state used for first time element is identified with crack	Fatigue damage exists which warrants analysis of the element to ascertain the serviceability of the element or bridge

Table 1.
Pontis system condition states related to fatigue [19].

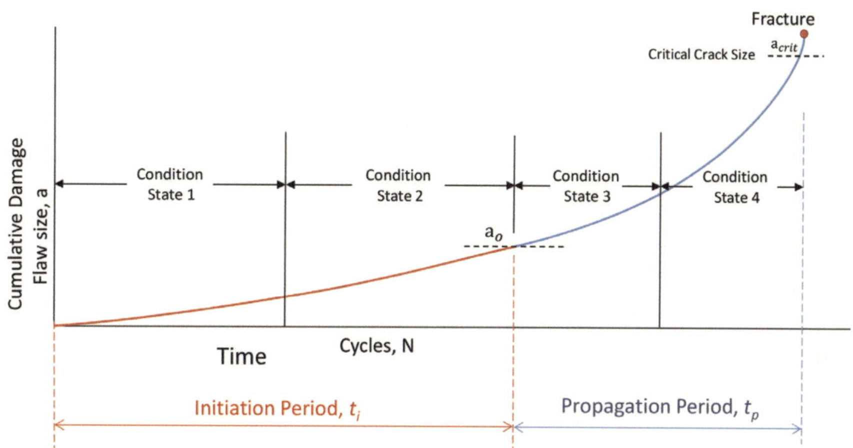

Figure 2.
BME condition states integrated into fatigue life curve.

hosts the commonly employed feasible actions that inspectors and state DOTs should take, given the condition state of their bridge. The purpose of the FHWA Bridge Preservation Guide is to provide a framework for a preventive maintenance program for bridge owners or agencies [20].

4. Case study

The fatigue assessment in this paper was conducted as part of the University of Maryland project to design and implement an integrated structural health monitoring system that is particularly suited for fatigue detection on highway bridges. Data for the analyses was acquired from a highway bridge carrying traffic from interstate 270 (I-270) over Middlebrook Road in Germantown, MD, seen in **Figure 3**. This bridge is referred to as the Middlebrook Bridge.

The Middlebrook Bridge was built in 1980 and reconstructed in 1991. With help from Maryland bridge inspectors, this bridge was selected as a good candidate for fatigue monitoring due to the average daily truck traffic, the bridge's maintenance

Figure 3.
Maryland bridge carrying I-270 over Middlebrook Road.

history, the geometric configuration, and the identification of existing fatigue cracks on the connection plates.

The Middlebrook Bridge is a composite steel I-girder bridge consisting of 17 welded steel plate girders with a span length of 140 ft. The bridge has three traffic lanes in the southbound roadway and five traffic lanes in the northbound, i.e., a high occupancy vehicle lane, an exit lane, and three travel lanes. Four fatigue cracks were reported in the Maryland State Highway June 2011 Bridge Inspection Report. These four cracks were all found in the welded connections between the lower end of the cross brace connection plate and the girder bottom flange.

The Middlebrook Bridge is built with skewed supports to accommodate the roadway below the bridge. Due to the skewed supports, the corresponding cross frames are also built with skewed angles. The Middlebrook Bridge was built with K-brace cross frame, seen in **Figure 4**.

The skew angle of the cross frames are built to code and are in accordance with AASHTO LRFD Bridge Design Specifications [11], so long as the skew angle is less than 20 degrees. A bridge with skewed cross braces is more prone to fatigue damages because its geometric configuration enhances the live load effects. The connections of the skewed cross braces are bent at an angle to connect with the transverse stiffeners of the bridge girders. When the bridge girders deflect, this angle introduces a bending effect into the transverse stiffeners.

Figure 4.
K-type cross brace on Middlebrook Bridge.

4.1 Structural health monitoring and data processing

A connection plate of a steel girder highway bridge is selected for long-term monitoring, shown in **Figure 5**.

This connection plate was identified by Maryland State Bridge inspectors in 2011 to have an existing active crack, i.e., a crack that is growing in size. The crack was described in inspection reports as "... very fine crack in the weld that connects the web stiffener to the top of the lower flange. The crack runs along the top of the weld material next to the stiffener and begins at the toe of the weld" [21].

Only one strain transducer was used to continue monitoring the bridge in a long-term monitoring evaluation. The strain transducer was placed on one of the stiffeners that showed to high tension stress. The bridge itself is loaded in bending by the dynamic effects caused from the vehicle passage. Specifically, **Figure 6** displays a sample of the acquired stress data as a function of time that was taken from a connection plate. The variation in loading of the load spectrum on the connection plate is dependent on the number of vehicles passing the bridge and the weight of the vehicle. Given that the traffic volumes and patterns are sporadic, the captured bridge loads are also sporadic. Strain data was collected from the bridge over the course of 1 year.

4.2 Fatigue analysis

The acquired variable amplitude strain data is converted to stress for linear damage accumulation models, where stress ranges are the main contributor to fatigue damage. In addition, methods of extrapolation were used to fill in missing points of data. The method of extrapolation that has been applied to the fatigue data is done in the rainflow domain. The results of the extrapolated rainflow matrix were modeled from a measured rainflow history, where the density of rainflow cycles was calculated. The calculation of this density provided the number of stress cycles and stress ranges that were to be estimated for each specific hour of the day. The data was then processed with the rainflow cycle counting method to count the number of stress ranges. **Figure 7** displays a histogram of measured stress ranges. This particular histogram displays the traffic data that was accumulated on the bridge over 8 days.

With variable amplitude stress history, the variable stress cycles are associated with a particular stress range value that will map the measured data with the S-N

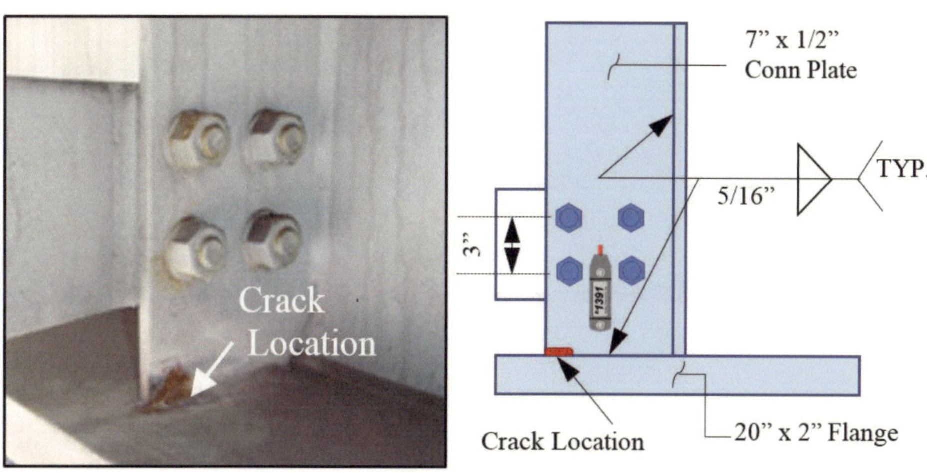

Figure 5.
Connection plate with known crack (left) and schematic of strain gage location (right).

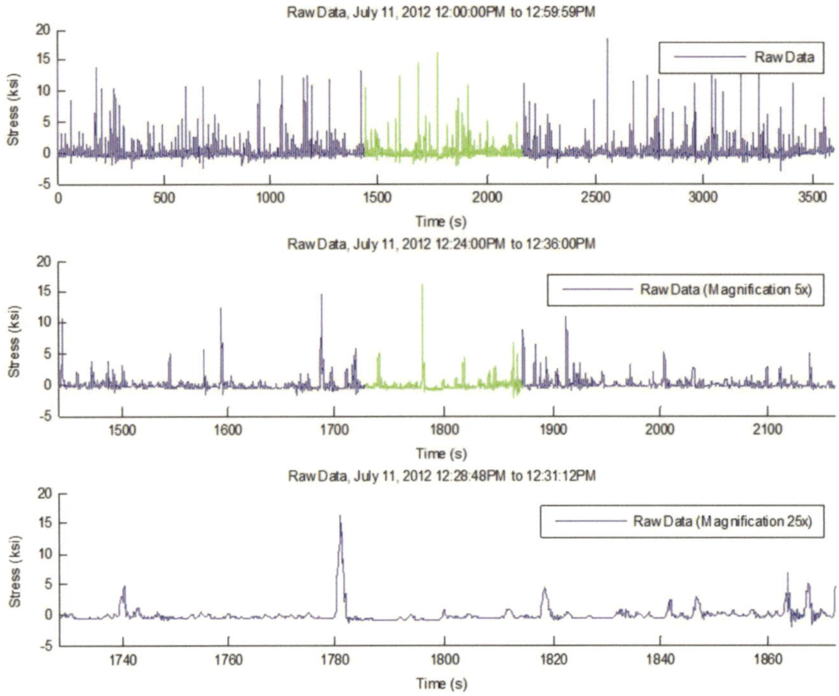

Figure 6.
Illustration of variable amplitude loading.

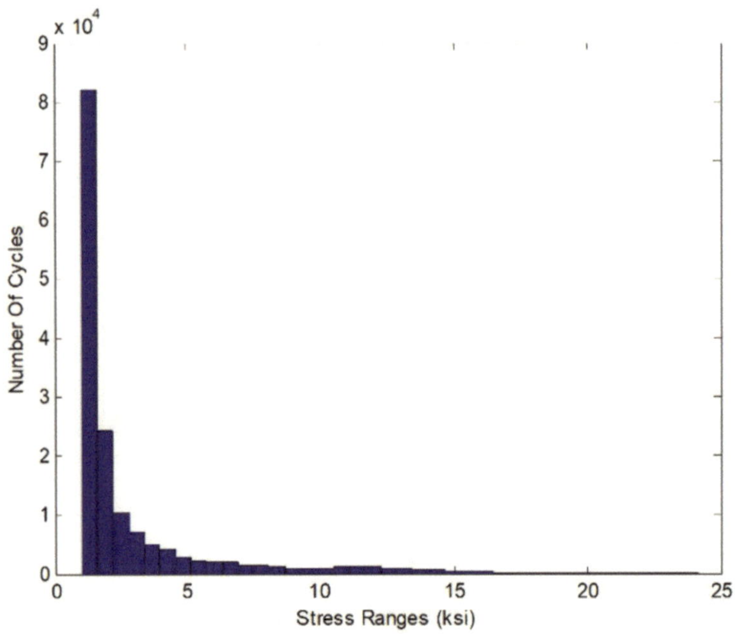

Figure 7.
Histogram of measured stress ranges.

curves. The measured histograms showed an un-proportionally large amount of cycles occur at smaller stress ranges. Therefore the stress ranges are truncated, and an effective stress range is solved for; with S-N curves the number of cycles to failure is based on the effective stress range. For this case study, the effective stress range (S_{re}) was found to be = 7.2 *ksi*, and the number of cycles over the course of 1 year were approximately 5.8 million cycles.

In accordance with the histograms for this case study, as the effective stress range increases, the number of stress cycles decreases dramatically. Without including an increase in traffic volumes, the effective stress range and number of cycles are assumed consistent for each year. Under this assumption, the estimated fatigue life for the crack initiation period was 18.0 years. **Figure 8** displays the yearly accumulation until failure is reached on the S-N curve.

4.3 Global model and simulation

A three-dimensional global model of the southbound direction, seen in **Figure 9**, was created to evaluate a bridge's response to loading. The model of the southbound superstructure consisted of eight I-girders. The concrete deck, the eight I-girders, and connection plates which connected cross frames to the girders were modeled by shell elements, while all the cross frames were modeled by spatial frames along their center of gravity. Special link members were defined to connect girder elements and concrete deck elements at the actual spatial points where these members intersect. The translations in the x-, y-, and z-directions were fixed at the abutments to represent the actual characteristics of support and continuity.

To study the dynamic effects of the Middlebrook Bridge, simulated truckloads were applied to the global finite element model through traffic simulation software, Traffic Software Integrated System (TSIS) 6.0. The data that was used to simulate the truckloads were taken from Maryland State Highway Administration's Internet Traffic Monitoring System (ITMS) and a local weigh station that is approximately 10 miles north of the Middlebrook Bridge but on the same interstate [23]. The ITSM features permanent Automatic Traffic Recorders that count traffic continuously throughout the year and breaks down the traffic count data by class, volume, and lane distribution [24]. The average hourly volume varied from 505 to 4215, and the

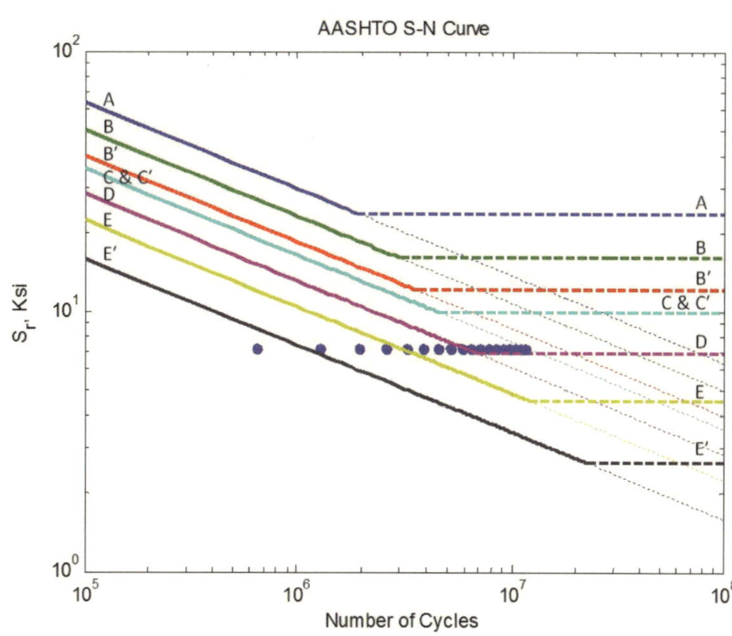

Figure 8.
AASHTO S-N curve with cumulative points plotted until failure.

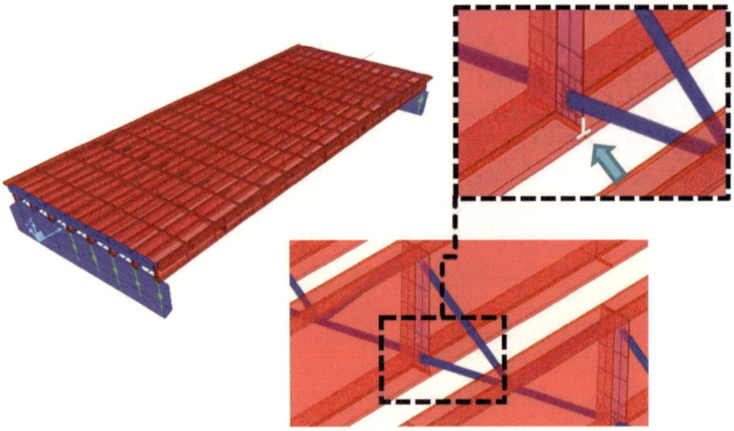

Figure 9.
Global model of Middlebrook Bridge and location of local model [22].

truck percentages also varied from about 10.5 to 20%. The weigh station-collected weight data of the truck traffic and the trucks were categorized into seven classes based on the number of axles. The majority of trucks were 2-axle which made up 25% of trucks and 5-axle, which made up 68% of trucks. The simulated truck network contained the mainline section of the highway with the Middlebrook Bridge in the center and adjacent ramps. Three classes of trucks were used for the simulation, shown in **Figure 10**. From the collected data, the simulation included the axle weight, axle spacing, vehicle position, and speed at each time step in the simulation.

The loading data from the simulation matched the loading data from field monitoring, and the simulated truckloads on the global model of the Middlebrook Bridge confirmed high tensile stresses between cross-frame connection plates and girder bottom flanges. These stresses are highest at the outer edge of the connection plate where the existing fatigue crack on the I-270 Bridge over Middlebrook Road was located. More detailed traffic load simulation is reported in a separate companion paper, *Fatigue Assessment of Highway Bridges Under Traffic Loading Using Microscopic Traffic Simulation*.

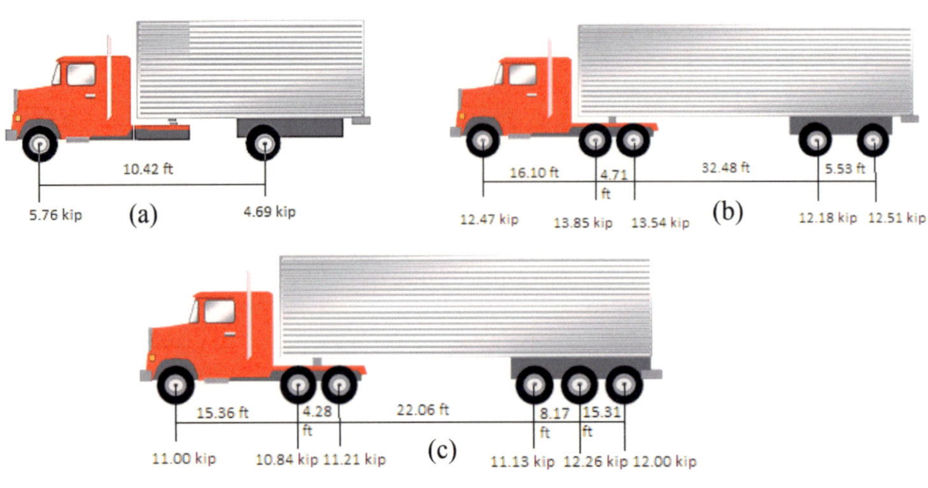

Figure 10.
Fatigue truck configurations (a), small truck, (b) medium truck, and (c) large truck.

4.4 Fracture analysis

Since the interest is to obtain a SIF, the global model cannot be any more refined, and a local model of this critical region was created for the purpose of understanding the stress field around the crack. A local model was created by applying the resulting deflections from the global model as resulting displacements in the local model. Since the deflections are a result of simulated traffic loads, applying the deflections simulates the loads transferred across a free-body section of the global model where the local model resides.

Additionally, the stress loads at the location of the strain gage were applied to the local model at the corresponding perimeter location. **Figure 9** displays the location of the local model within the global structure. This location is described with white lines that outline the local model geometry. **Figure 11** displays the local model with applied displacements and forces. A dashed rectangle outlines the location of the existing crack. A fine mesh is created around the previously identified existing crack, and a radial mesh is created around the crack tip. The crack was modeled with an assumed depth of 0.05 inch, which is slightly greater than a largest depth of micro-crack (0.05 mm <a <1 mm) and approximately the length of the penetration of the fusion in a fillet weld [25]. **Figure 12** displays the stress contour of the y-component of the cross section and a magnified view at the location of the crack.

4.4.1 Damage tolerance and fracture toughness

The specifications of the American Society for Testing and Materials for A572 Grade 50 steel require a minimum yield strength value of 50 ksi. The fracture toughness for the steel on the Middlebrook Bridge is $K_{IC} = 56\, ksi\, \sqrt{in}$. The critical crack length that corresponds to the fracture toughness comes from the fracture mechanics equation for critical SIF. Under the parameters that fit the Middlebrook Bridge, the critical crack size is $a_{crit} = \frac{K_{IC}}{\pi \beta^2 \sigma^2} = .15\, in$.

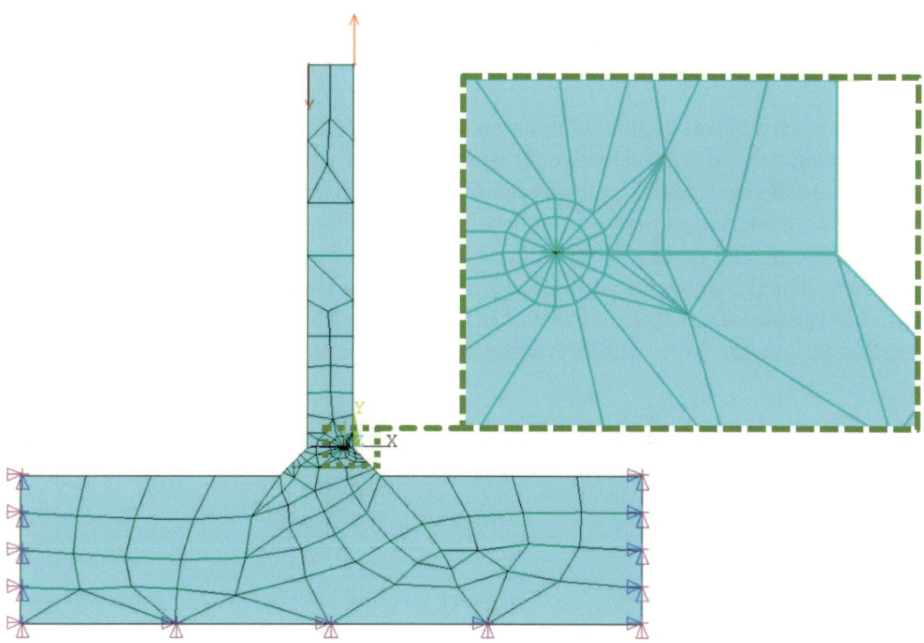

Figure 11.
FEM local model with applied displacements and forces.

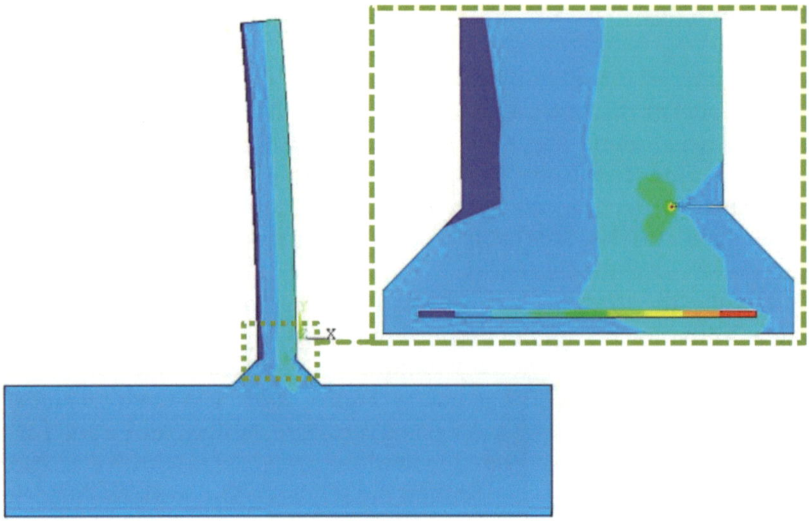

Figure 12.
Stress contour of crack to illustrate plastic zone at crack tip.

4.4.2 Crack growth and total cumulative damage

The computed SIF from the local model was used alongside Paris law to solve for the yearly crack growth rate. Rearranging Eq. (5), the crack size, a, at any given time, is a function of the SIF and the effective stress range. The accumulation of damage for fatigue crack growth models (shown in Eq. 6) is consequent of the change in crack size, Δa_i; then the crack would reach the critical size after 9.6 years. Since the bridge inspectors first noticed the bridge cracking in 2011, at the time of testing (2012–2013), the crack had been present for about 1–2 years. The crack was repaired in 2014, at which time the remaining useful life for this bridge element was calculated to be 6.6 years to failure.

4.5 Integration of damage with Maryland condition states

The case study was estimated from measured and extrapolated load distributions to assess the life of the bridge. The fatigue life of the crack initiation period was found to be 18 years, and the fatigue life of the crack propagation period was found to be about 9 years. Accordingly, rate adjustment factors were selected to be $\alpha_I = 0.7$ and $\alpha_P = 0.3$. The second row in **Table 2** illustrates the amount of damage for each condition state. The third row is a simplified explanation of the condition states which are found in the *Maryland Pontis Element Data Collection Manual*, and the last row is the feasible actions for these condition states from the FHWA Bridge Preservation Guide.

	Condition state 1	Condition state 2	Condition state 3	Condition state 4
D_{Total}	$0 \rightarrow \frac{1}{2}\alpha_i$	$\frac{1}{2}\alpha_i \rightarrow \alpha_i$	$\alpha_i \rightarrow \left(\alpha_i + \frac{1}{2}\alpha_p\right)$	$\left(\alpha_i + \frac{1}{2}\alpha_p\right) \rightarrow (\alpha_i + \alpha_p)$
D_{Total}, %	0–35%	35–70%	70–85%	85–100%
Cracking/fatigue	None	Fatigue damage	Analysis warranted	Severe fatigue damage
Feasible action	Do nothing	Preventive maintenance	Rehabilitation	Rehabilitation or replacement

Table 2.
Damage accumulation mapped to bridge condition states.

In 2014, when the crack was repaired, the calculated percent damage was 87.2%, correlating to condition state 4, "Fatigue damage exists which warrants analysis of the element to ascertain the serviceability of the element or bridge."

5. Summary and conclusions

This paper proposes a damage accumulation model to more accurately characterize fatigue damage prognoses of bridge elements. The fatigue life has been described and divided into two periods: the initiation period and the propagation period. An empirical correlation approach, characterized by the S-N curve, is used to analyze the initiation period, and the data acquired from SHM and traffic simulation models are used to inform the crack initiation analyses. SHM is shown to have a significant contribution in damage prognosis, where the sensing information instrumentation is used to validate FEM models and acquire information about a bridge's response to loads. It is shown how this data can be particularly useful when processed through cycle counting algorithms, and methods of extrapolation are applied to gather information on stress range distributions to estimate future traffic loads of the bridge. Fatigue damage assessments in the crack initiation period can be supplemented with a fracture mechanics analysis, which defines the crack propagation period and estimates crack growth. It is also shown how finite element modeling can be used to solve for the SIF, which is then used to estimate the growth rate. A case study is presented to illustrate the application of the fatigue damage prognoses on a steel highway bridge element. The damage accumulation models are used to estimate the onset of a fatigue crack and fatigue crack growth rates and ultimately derive a damage prognosis of the bridge element.

Acknowledgements

This research was partially sponsored by the US Department of Transportation's Office of the Assistant Secretary for Research and Technology (USDOT/OST-R), under The Commercial Remote Sensing and Spatial Information (CRS&SI) Technologies Program. This support is acknowledged and greatly appreciated.

Author details

Timothy Saad[1], Chung C. Fu[1]*, Gengwen Zhao[2] and Chaoran Xu[1]

1 The Bridge Engineering Software and Technology (BEST) Center, Department of Civil and Environmental Engineering, University of Maryland, College Park, MD, USA

2 Virginia Department of Transportation, Richmond, VA, USA

*Address all correspondence to: ccfu@umd.edu

References

[1] FHWA. Focus Accelerating Infrastructure Innovations. Federal Highway Administration Launches Steel Bridge Testing Program. 2011

[2] Haldipur P, Jalinoos F. Detection and Characterization of Fatigue Cracks in Steel Bridges. 2010. Available from: www.structuralfaultsandrepair.com

[3] FHWA. Bridge Inspector's Reference Manual. Vol. 2. Washington DC: FHWA NHI Publication No. 12-050; 2012

[4] Schijve J. Fatigue of Structures and Materials. 2nd ed. Netherlands: Springer Science+Business Media B.V.; 2009

[5] Mohammadi J, Guralnick SA, Polepeddi R. Bridge fatigue life estimation from field data. Practice Periodical on Structural Design and Construction. 1998;**3**(3):128-133

[6] Zhou YL, Maia NM, Sampaio RP, Wahab MA. Structural damage detection using transmissibility together with hierarchical clustering analysis and similarity measure. Structural Health Monitoring. Sage Publication. 2017;**16**(6):711-731. Available from: https://www.nafems.org/downloads/FENet_Meetings/Trieste_Italy_Sep_2002/FENET_Trieste_Sept2002_DLE_Zafosnik.pdf/

[7] Shantz CR. Uncertainty Quantification in Crack Growth Modeling Under Multi-Axial Variable Amplitude Loading. Nashville, Tennessee: Graduate School of Vanderbilt University; 2010

[8] ASTM E-1049. Standard Practices for Cycle Counting in Fatigue Analysis. West Conshohocken: ASTM International; 2011

[9] AASHTO. Standard Specifications for Highway Bridges. Washington, D.C.: American Association of State Highway and Transportation Officials. 2002

[10] NCHRP. Fatigue Evaluation of Steel Bridges. Washington, D.C.: National Academy of Sciences, Transportation Research Board. 2012

[11] AASHTO. LRFD Bridge Design Specifications. 7th ed. Washington DC: American Association of State Highway and Transportation Officials. 2014

[12] Massarelli PJ, Baber TT. Fatigue Reliability of Steel Highway Bridge Details. US DOT FHWA, Charlottesville, Virginia: Virginia Transportation Research Council; Virginia DOT. 2001

[13] AASHTO. Guide Specifications for Fatigue Evaluation of Existing Steel Bridges. Washington DC: American Association of State Highway and Transportation Officials; 1990

[14] Zhou YE. Assessment of bridge remaining fatigue life through field strain measurement. Journal of Bridge Engineering. 2006;**11**(6):737-744

[15] Keating PB, Fisher JW. Fatigue Tests and Design Criteria. Bethlehem, PA: National Cooperative Highway Research Program and Fritz Engineering Laboratory; 1986

[16] Zafosnik B, Ren Z, Ulbin M, Flasker J. Evaluation of stress intensity factors using finite elements. FENet: A NAFEMS Project; 2002

[17] Mertz D. Steel Bridge Design Handbook: Design for Fatigue. Washington, D.C.: FHWA-IF-12-052-Vol.12; 2012

[18] CAE Associates. Fracture Mechanics in Workbench v14.5 ANSYS e-Learning Session. Middlebury. 2013

[19] MDSHA. Pontis Element Data Collection Manual. Baltimore, MD: Bridge Inspection and Remedial Engineering Division, Office of Bridge Development; 2003

[20] FHWA. Bridge Preservation Guide. U.S. Department of Transportation Federal Highway Administration. New Jersey Avenue, SE Washington, DC. 2011

[21] MDSHA. Maryland State Highway Administration Bridge Inspection Report. MDSHA, Maryland State Highway Administration. North Calvert Street, Baltimore, Maryland. 2013

[22] Fu CC, Wang S. Computational Analysis and Design of Bridge Structures. Boca Raton, FL: CRC Press Taylor and Francis Group; 2014

[23] SHA. Internet Traffic Monitoring System. Maryland State Highway Administration. 29 July 2015. [Online]. Available from: http://shagbhisdadt.mdot.state.md.us/ITMS_Public/default.aspx

[24] Zhao G. Truck Loading Simulation for the Fatigue Assessment of Steel Highway Bridges. College Park: University of Maryland; 2015

[25] Janosch J. Investigation into the Fatigue Strength of Fillet Welded Assemblies of E-36-4 Steel as a Function of the Penetration of the Weld Subjected to Tensile and Bending Loads. 1993

Chapter 5

Fatigue Assessment of Highway Bridges under Traffic Loading Using Microscopic Traffic Simulation

Gengwen Zhao, Chung C. Fu, Yang Lu and Timothy Saad

Abstract

Fatigue is a common failure mode of steel bridges induced by truck traffic. Despite the deterioration caused by environmental factors, the increasing truck traffic volume and weight pose a premier threat to steel highway bridges. Given the uncertainties of the complicated traffic loading and the complexity of the bridge structure, fatigue evaluation based on field measurements under actual traffic flow is recommended. As the quality and the quantity of the available long-term traffic monitoring data and information have been improved, methodologies have been developed to obtain more realistic vehicular live load traffic. A case study of a steel interstate highway bridge using microscopic traffic simulation is presented herein. The knowledge of actual traffic loading may reduce the uncertainty involved in the evaluation of the load-carrying capacity, estimation of the rate of deterioration, and prediction of remaining fatigue life. This chapter demonstrates a systematic approach using traffic simulation and bridge health monitoring-based fatigue assessment.

Keywords: fatigue, finite element modeling, truck traffic loading, microscopic traffic simulation, cross-frame

1. Introduction

Fatigue is a common failure mode of steel bridges. About 80–90% of failures in steel structures are related to fatigue and fracture [1]. Despite the deterioration caused by environmental factors, the increasing traffic volume and weight pose a premier threat to steel highway bridges. The total number of truck passages in the 75 year life of a highway bridge could exceed 100 million [2]. With the aging of existing steel highway bridges and the accumulated damage under truck loading, the fatigue assessment for continuing service has become important for decision makings on the structure maintenance, component replacement, and other major retrofits.

Given the uncertainties of the complicated traffic loading and the complexity of the bridge structure, fatigue evaluation based on field measurements under actual traffic flow is recommended by many researchers. As the quality and quantity of the available long-term traffic monitoring data and information have been improved, a set of methodologies has been developed to obtain a more realistic vehicular live load. The knowledge of actual traffic loading may reduce the uncertainty involved in the evaluation of the load-carrying capacity, estimation of the rate of deterioration, and prediction of remaining fatigue life. However, there are still some difficulties in field measurements. For example, some highway bridges are not accessible for field tests; the maintenance of monitoring system is difficult and costly, especially for long-term monitoring; some highway bridges will not even be considered for field tests with economic concerns.

The results of several NCHRP reports, written by Dr. John Fisher in 1970s, have confirmed that for welded details, fatigue life is primarily a function of stress range, detail category, and the number of applied cycles. The live load of a bridge includes static and dynamic parts while the early research and studies focused on the static portion. Schilling [3] and Raju et al. [4] suggested to improve the accuracy of the fatigue truck model by adjusting the fatigue truck axle weights in proportion to an equivalent total weight calculated from the specific site load distribution. The collected weigh station measurements, or data measured in stationary weight scales, were used by Nowak et al. [5] to determine the truck-load spectra for highway bridges on highways I-75 and I-94. Later, Laman and Nowak [6] developed a fatigue-load model from weigh station measurements and calculated the statistical parameters of stress for girder bridges. The results indicated that magnitude and frequency of truck load spectra are strongly site-specific and the live load stress spectra are strongly component-specific. With the advantage of weigh-in-motion (WIM) technology, Miao and Chan developed a methodology by using 10 year WIM data for deriving highway bridge live load models for short span bridges in Hong Kong [7]. NCHRP developed a set of protocols and methodologies for using available nation-wide, state-specific, or site-specific truck traffic data collected at different U.S. sites to obtain live load models for LRFD superstructure design, fatigue design, deck design, and design for overload permits [8].

In the early studies, it was commonly assumed that a certain percentage of the total weight was loaded on the front axle or rear axle for the magnitude and configuration. Further, there was no real traffic simulation considering the truck flow pattern. Bridge behavior simulations under truck loading were usually performed using the Monte Carlo method, which is a statistical projection approach with generic nature and does not consider any vehicle and driver behavior models when simulating truck traffic flow. In recent years, traffic flow simulation method has been applied to provide instantaneous information of individual vehicle by many researchers. Chen and Wu developed a general framework of modeling the live load from traffic for a long-span bridge by using the cellular automation (CA) traffic flow simulation technique. A typical four-lane long-span bridge was studied using the proposed method. Each lane was divided into cells with an equal length of 7.5 m. Three conditions, the free flow, the moderate flow, and the congested flow, were considered in the simulation. A simple comparison between the simulated static traffic load and the AASHTO LRFD HL-93 design load was conducted. The results showed that the HL-93 may be insufficient for the congested flow condition [9].

This research has developed a framework for the fatigue assessment of steel highway bridges based on simulated truck loading. The proposed methodology is implemented on a steel highway girder bridge on interstate 270 (I-270) over Middlebrook Road in Germantown, Maryland. With the help of the available long-term monitoring traffic data, truck loading was also obtained through the probability-based model. Then, the three-dimensional finite element (FE) global bridge models were studied subjected to the simulated truck loading. Meanwhile, the preliminary field test and the long-term monitoring test were also conducted. The FE models were calibrated with the collected field measurements through monitoring systems, and the simulated numerical structural responses were validated.

Lastly, this model has been used for identifying the cause of fatigue cracks reported in the biennial bridge inspection. Thus, the proposed methodology could be used to realistically simulate the fatigue behavior of steel highway bridges under current or future truck loading, to direct the experimental designs and instrumentation plans before performing experiments on laboratory or on site, and to better understand the fatigue mechanism and prevent the fatigue damage of steel highway bridges.

2. Fatigue cracks and bridge testing

2.1 Bridge introduction

The I-270 Bridge over Middlebrook road (MD Bridge No.15042) is a simple-span composite steel I-girder bridge with a span length of 140 ft. This bridge is comprised of two structures for the northbound (NB) and southbound (SB) roadways respectively, separated at the centerline. It carries three traffic lanes in the southbound and four traffic lanes in the northbound with equal lane widths of 12′-0″.

The southbound superstructure provides a curb-to-curb roadway width of 61′-2″ and consists of eight identical welded steel plate girders with a composite reinforced concrete deck constructed with shear connectors. The eight girders are equally spaced at 7′-11″ and each girder has a constant web depth of 60″ throughout the entire bridge. The northbound superstructure provides a curb-to-curb roadway width of 73′-1″ and consists of nine identical welded steel plate girders with a composite reinforced concrete deck constructed with shear connector. The nine girders are equally spaced at 8′-5″ and each girder has a constant web depth of 60″ throughout the entire bridge. This bridge has a 76 degree parallel skew of its bearing lines (or 14 degree measured from normal). The cross-frames are inverted K-type braces with bottom chords only. All of them are parallel to the bearing lines. Girders of the southbound superstructure are numbered as G1 through G8 from the exterior to the centerline of the bridge. The cross section is depicted in **Figure 1**.

Designed in 1988, the I-270 Bridge over Middlebrook Road has been in-service for over 20 years. In addition to the deterioration caused by environmental factors, the bridge structure has also been subjected to increasing traffic volume and weight. Four fatigue cracks as marked on **Figure 2** were reported in the June 2011 Bridge Inspection Report, and all in the welded connection between the lower end of the cross frame (**Figure 3**) connection plate and the girder bottom flange of the southbound superstructure. **Figure 4(a)** shows one of the four crack locations at G3B2D3 (Girder 3 Bay 2 Diaphragm 3). Therefore, only the southbound superstructure will be discussed in the following sections. Most bridge components with fatigue cracks are repaired or replaced shortly after the crack is found in an inspection. However, since the crack on the I-270 Bridge was identified on a secondary bridge member,

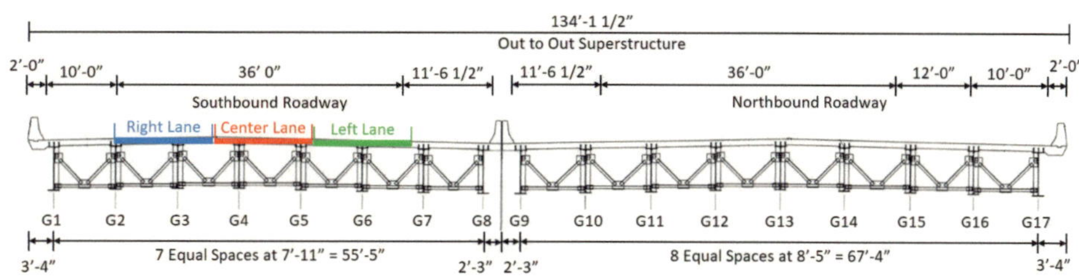

Figure 1.
Cross section with lane positions.

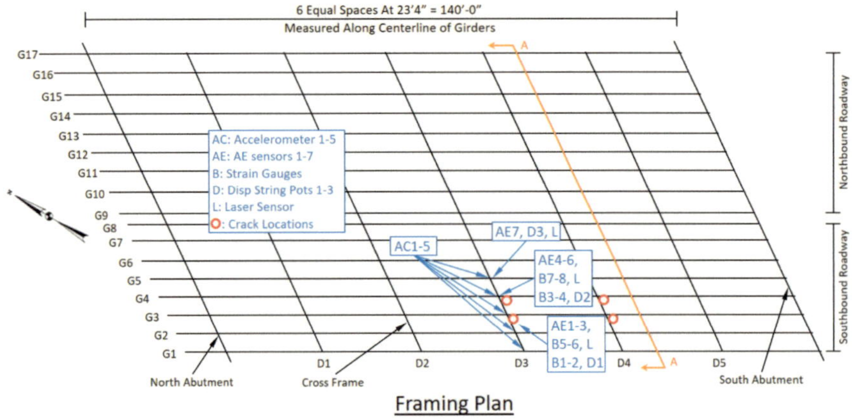

Figure 2.
Crack locations and sensor placements on the framing plan.

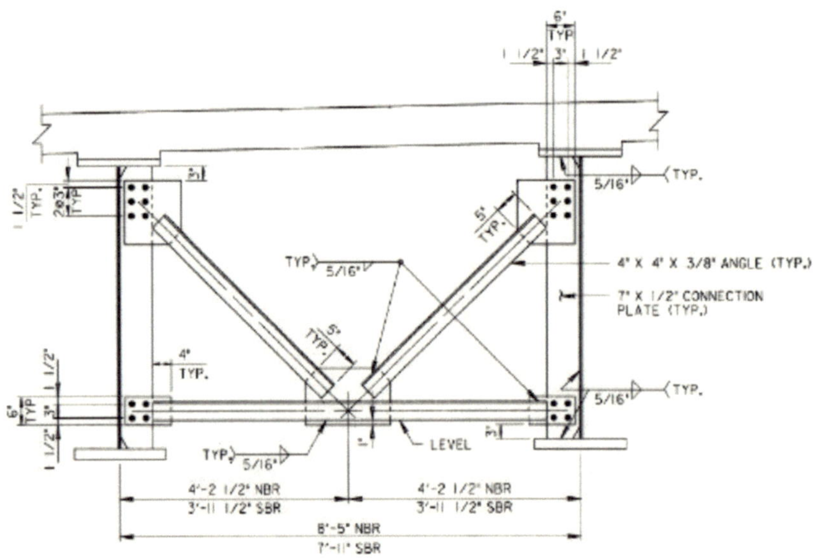

Figure 3.
Cross frame detail.

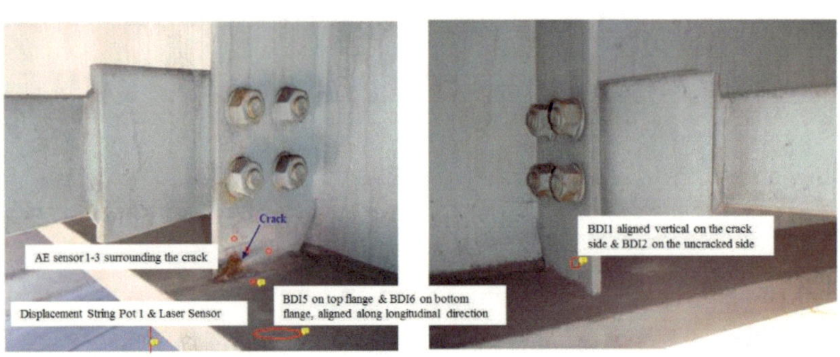

Figure 4.
Crack locations and sensor placements: (a) details at G3B2D3 and (b) details at G3B3D3.

and delaying repair would not jeopardize the safety of drivers, this crack was selected for research purposes and long-term monitoring.

2.2 Field test and results

The field test of the I-270 Bridge was conducted through a Wireless Integrated Structural Health Monitoring System sponsored by the U.S. Department of Transportation, Office of the Assistant Secretary for Research and Technology (USDOT/OST-R). This smart bridge condition monitoring system, termed the ISHM system, features a number of technology innovations, including remote sensing capability, piezo paint acoustic emission sensors, wind and solar based energy harvesting devices to power the sensor network, high-speed wireless sensing ability and advanced data analysis methods for remaining life estimation of aging bridges. Through successful advancement and commercialization in the state-of-the-art technology of remote infrastructure sensing, the ISHM system is promising to reduce life cycle costs while significantly maintaining the sustainability of the highway infrastructure in the US.

2.2.1 Instrumentation plan

The main data acquisition systems used in this test consisted of a PXI-based data acquisition system by National Instruments, which was used for data collection by the BDI strain transducers, string pots and Acoustic Emission (AE) sensors, and a multi-channel data acquisition equipment CR5000 manufactured by Campbell Scientific, Inc., which was used for the extra BDI strain transducer. Types of sensors used in this project were: (1) piezoelectric paint AE sensors; (2) wireless accelerometers; (3) laser sensor; (4) ultrasonic distance sensors; (5) BDI strain transducers; and (6) string pots. Sensors were strategically placed where the cracks on the SB bridge were identified and their related strain, AE or supplemental data can be collected by the data acquisition system and later used for validating the FE models. The instrumentation plan is shown in **Figure 2**. Girder displacement and stress range records due to truck traffic were part of the field measurements in this study.

2.2.2 Vibration response

A total of four wireless accelerometers were used to monitor the vibration responses of the bridge. Wireless sensors were installed on four girders (Girders 2–5) and acceleration data were acquired at 100 Hz sampling rate synchronically. The acceleration data were used to provide modal frequency information that was used to calibrate the finite element model of the bridge. The fundamental frequency measured is 3.22 Hz, which was very close to the value of the first vertical mode from the finite element analysis of 3.24 Hz discussed in the following sections.

2.2.3 Bridge deflection

Both laser sensor and ultrasonic distance sensors were used to measure the dynamic deflection of the bridge. Only one laser sensor and one ultrasonic distance sensor were used each time. The measured deflection value from the laser sensor agreed well with the string pot, and its accuracy was also validated by the fundamental frequency indicated by fast Fourier transform (FFT) of the laser sensor measured deflection data. The measured maximum deflection of the I-270 bridge over Middlebrook Road under traffic loading is summarized in **Table 1**.

Girder number	3	4	5
Max D (in)	0.2598	0.2717	0.2480

Table 1.
Maximum deflections measured by laser sensor.

2.2.4 Stress

Cracks occur in the direction perpendicular to the direction of principal tensile stress. To assess the driving force of the fatigue cracks in the connection welds, strain gages were placed vertically on the connection plate just beyond the tip of the existing crack. Strain gages were also placed longitudinally on the girder flanges to correlate with the occurrence of vehicular loads. For comparison with the results from analytical methods, field testing is applied as it is the most accurate approach since no assumptions need to be made for uncertainties in load distribution such as unintended composite action between structural components, contribution of nonstructural members, stiffness of various connections, and behavior of the concrete deck in tension. The actual strain histories experienced by bridge components are directly measured by strain gages at the areas of concern. The effects of varying vehicle weights and their random combinations in multiple lanes are also reflected in the measured strains.

BDI 1-4 strain transducers were placed on both sides of the connection plates while BDI 5-8 strain transducers were placed at the top and bottom flanges on Girders 3 and 4 (**Figure 2**). **Figure 5** shows the measured stresses on the flanges and connection plates, respectively. The maximum stress measured at the bottom flange was 1.604 ksi in tension for BDI 3215 on the bottom flange of Girder 3 due to regular traffic loading, which was very low comparatively. As for the connection plates, the maximum stresses were 16.18 ksi in tension for BDI 1641 on Girder 3 and 16.1 ksi in tension for BDI 1644 on Girder 4 (**Figure 5**).

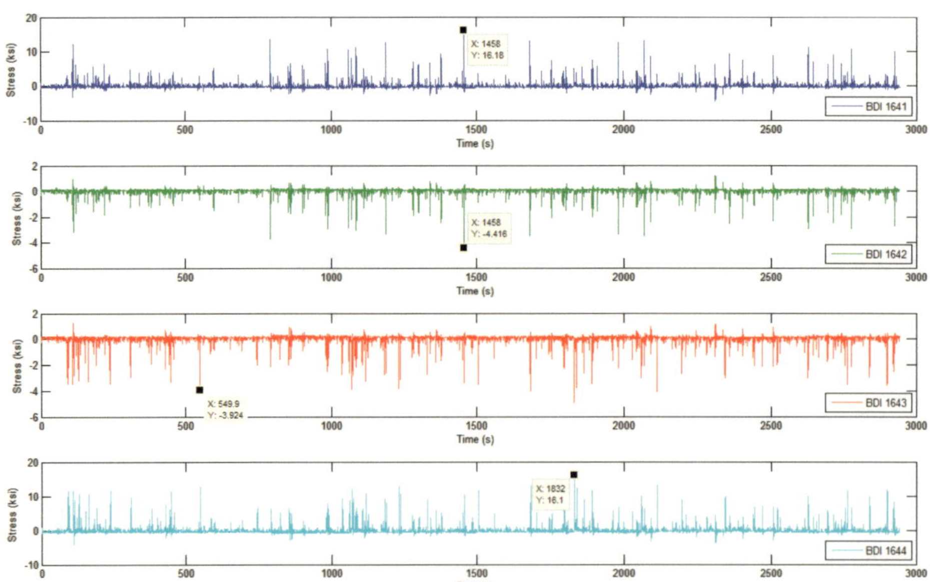

Figure 5.
BDI strain transducer measurements of connection plates and flanges (positive indicates compression; 1641 G3 cracked side; 1642 G3 uncracked side, 1643 G4 uncracked side and 1644 G4 cracked side).

3. Finite element modeling

3.1 Traffic loading

The traffic data that is used to simulate the traffic flow were the time-varying vehicle count data obtained from the Traffic Monitoring System Program (TMSP); operated and maintained by the Highway Information Services Division under Maryland Department of Transportation, State Highway Administration (MDSHA) [10]. The TMSP has been responsible for the collection, processing, analysis, and management of Maryland highway traffic data since 1997. Under this program, MDSHA has implemented 79 permanent continuous automatic traffic recorders (ATRs) counting traffic continuously throughout the year, and over 3800 short-term (48 hour) program count locations throughout the state, with data taken during the week on either Tuesday and Wednesday or Wednesday and Thursday to reflect typical weekday travel patterns. These monitoring systems are installed across Maryland and monitor most of the arterials, freeways, and interstates. **Figure 6** displays the location of several ATRs on I-270.

The TMSP has also created an Internet Traffic Monitoring System (I-TMS) that provides access to detailed traffic count data. On the I-TMS, the user can select an individual location to view reports (class, volume, lane distribution, etc.). Based on the hourly traffic volume, one typical day was divided into four different time periods: midnight, early morning and night, morning peak hour, and noon to evening, as shown in **Table 2**. The durations for these four time periods are 5, 5, 5, 9 hours, respectively. The average hourly volume varied from 505 to 4215, and the truck percentage also varied from 10.39 to 20.10%. Lane distribution of I-270 Bridge over Middlebrook Road is shown in **Table 3**. The main purpose of this time division is to realistically simulate the major characteristics of the traffic flow for each time period.

Following the specifications in the Guide Specifications for Fatigue Evaluation of Existing Steel Bridges (1989), weigh station measurements were collected at Hyattstown Weigh Station. From these measurements, a gross-weights histogram was obtained for the truck traffic, which was used to calculate the effective gross weight of the fatigue truck. The Hyattstown Weigh and Inspection Station is located approximately 10 miles north of I-270 Bridge over Middlebrook Road,

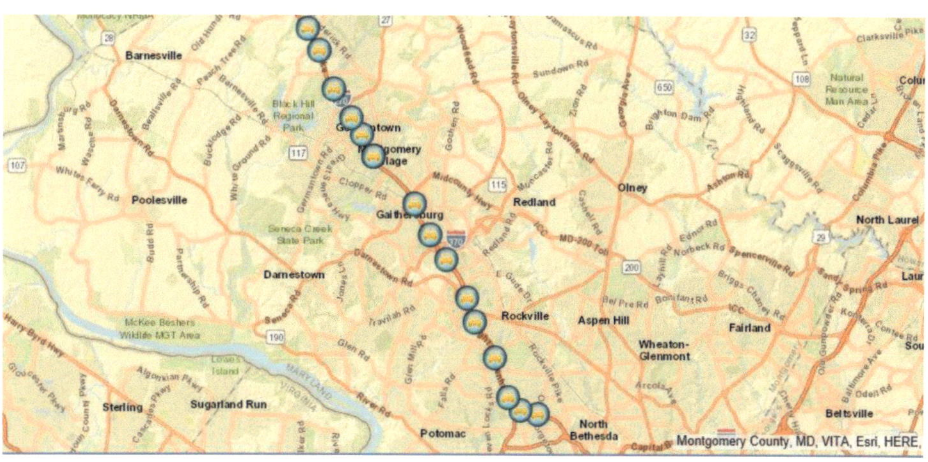

Figure 6.
Observation points on I-270 in Montgomery County.

Time period	Time	Average total volume (no. of vehicles per hour)	Passenger car (no. of vehicles per hour)	Truck by axle number (no. of vehicles per hour)					Truck percentage
				2	3	4	5	6	
Midnight	23:00–24:00 0:00–3:00 (5 hours)	505	403	25	2	3	69	3	20.10%
Early morning and night	4:00–5:00 19:00–23:00 (5 hours)	1934	1712	55	4	7	150	6	11.40%
Moring peak	5:00–10:00 (5 hours)	4215	3759	113	7	14	310	12	10.82%
Noon to evening	10:00–19:00 (9 hours)	3021	2707	78	5	9	213	8	10.39%

Table 2.
Traffic condition under different time period.

Vehicle type	Left lane (%)	Middle lane (%)	Right lane (%)
Total (passage car and truck)	31.87	30.62	37.51
Truck	1.45	44.84	53.71

Table 3.
Lane distribution of one typical day.

along Interstate 270 (I-270). Around 2200 samples during 1 year were chosen as the database to generate the truck weight and configuration. The measured data were filtered before the statistical analyses were made, where five samples were deleted. All the trucks were cataloged into seven classes based the number of axles (**Figure 7**). It is clear that 2-axle trucks and 5-axle trucks were the majority, which occupies about 24.87 and 67.99%, respectively. The 3-axle trucks, 4-axle trucks and the heaviest 6-axle trucks and over accounted for 1.61, 2.98 and 2.55%, respectively, which adds up to 7.14% in total.

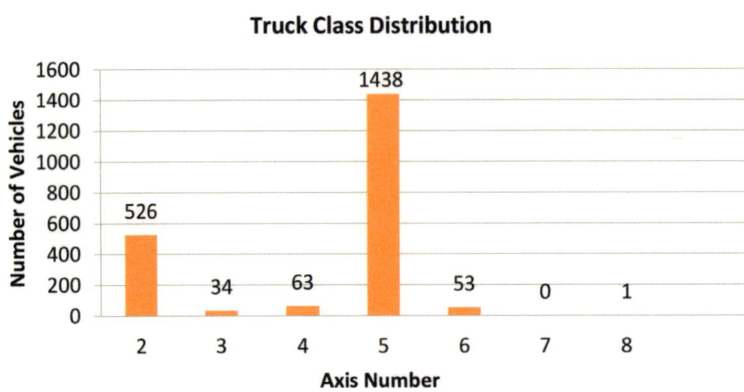

Figure 7.
Truck class distribution.

Because of the limitation of the traffic simulation software CORSIM, only three different types of trucks can be defined during traffic simulation: the small truck, the medium truck, and the large truck. Since 2-axle trucks and 5-axle trucks were the majority, the small truck was defined to consist 2-axle trucks and 3-axle trucks, and the medium truck to include 4-axle trucks and 5-axle trucks. For safety consideration, the heaviest 6-axle trucks and over were also considered as the third type, although it only takes a very small percentage.

The effective gross weight of the fatigue truck was computed from Eq. (1)

$$W = \left(\Sigma f_i W_i^3\right)^{1/3} \quad (1)$$

where f_i is the fraction of gross weights within an interval and W_i is the midwidth of the interval. The gross weight was distributed to axles in accordance with the site data. The final fatigue truck configurations were shown in **Figure 8**.

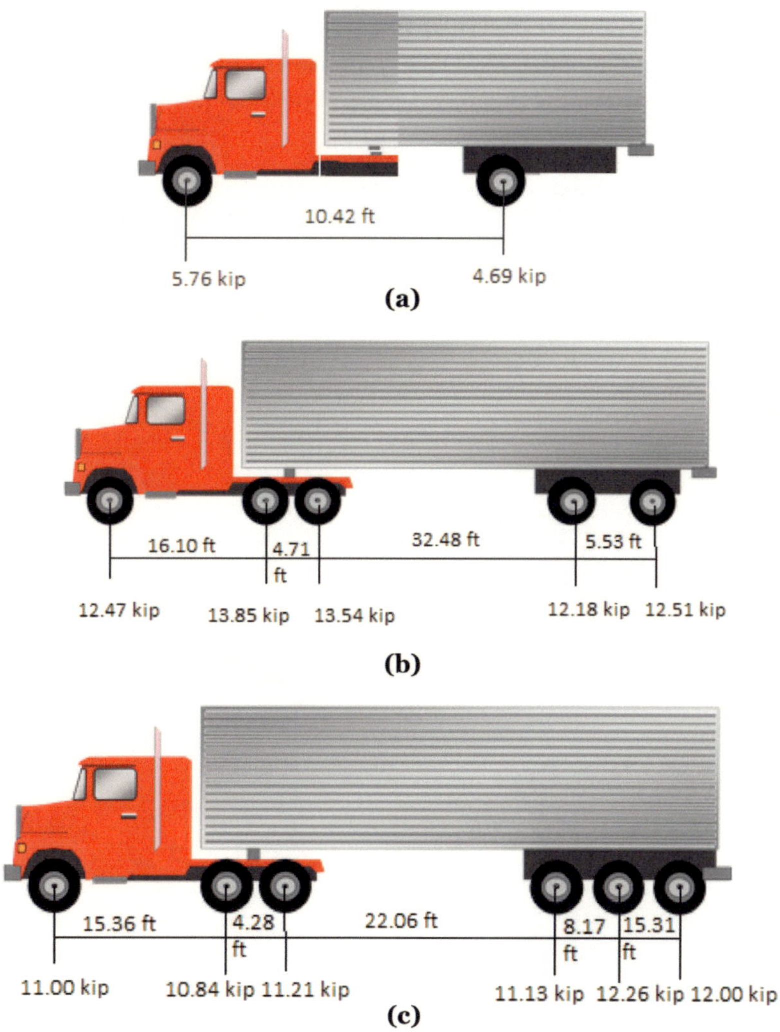

Figure 8.
Fatigue truck configurations: (a) small truck; (b) medium truck; and (c) large truck.

As an intercorrelated component of a whole transportation network, the actual traffic flow through a bridge is affected by the traffic on the connecting roadway segments. Therefore, to realistically capture the major characteristics of the traffic flow, a road network system consisting of the bridge, highway, and two neighboring ramps was studied in the present work. The detailed procedure is summarized in four steps:

1. Build the simulation network (**Figure 9**) in TSIS 5.1 [11] around the I-270 over Middlebrook Road based on the background map obtained from Google Maps. The background map is adjusted to the correct scale, and the simulation network is drawn along the real roadway segment. The network contained the mainline of I-270 and adjacent on-ramps of the bridge in the study. Since the focus is on the southbound of the bridge, the network only contains one-way southbound link. The simulation time is set to be 1 hour.

2. Use the time varying vehicle count data collected from nearby detectors, which were placed on the I-TMS website, and combine with the weigh station measurements collected from the Hyattstown southbound station as the input data for the simulation model. The truck count data from the vehicle count report are converted to truck percentage (truck count/total vehicle count) as the input for CORSIM simulation.

3. Set three different types of trucks corresponding to fatigue trucks generated in the last section. Install three loop detectors at the bridge in the created simulation network, one for each lane to record the speed, type, and passage time of the detected vehicles.

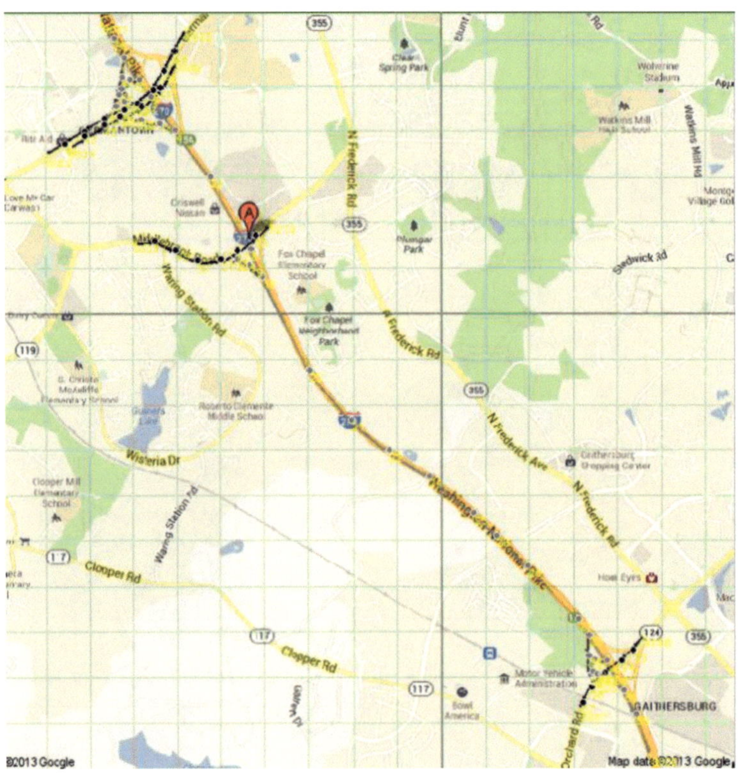

Figure 9.
Traffic simulation network.

4. After the simulation network is created, the traffic demand is input and calibrated, and the detectors are installed, the CORSIM simulation can begin. The simulation provides the following meaningful results for the analysis. First, it records the animation of the simulation, which is used to observe the passage time of the trucks. Second, it provides text output including the volume and speed statistics by each interval (set to be 1 s here). Combining the above two outputs, the passage time and the lane occurred and speed of the truck could be successfully matched.

The results (**Table 4**) could provide vehicle position and speed at each time step of the simulation. It was found from the results that the 20 min simulation period currently used can lead to stable pattern and matched field monitoring results. Details of the field monitoring are reported in a separate companion paper, *"Integrating Bridge Management Systems with Fatigue Damage Assessments."*

3.2 Bridge global model

To investigate the fatigue performance of the bridge, a three-dimensional finite element model was developed for linear-elastic structural analysis using the CSiBridge [12], as depicted in **Figure 10**. The model of the southbound superstructure consisted eight I-girders. The concrete deck, the eight I-girders, and connection plates which connected cross-frames to the girders were modeled by shell elements, while all the cross-frames were modeled by spatial frames along their center-of-gravity. Special link members were defined to connect girder elements and concrete deck elements at the actual spatial points where these members intersect. The translations in the x-, y-, z-directions were fixed at the abutments to represent the actual characteristics of support and continuity. It is complicated to establish a comprehensive finite element model of a large practical structure for fatigue damage analysis, since the finite element model should

Time periods	CORSIM		
	Average speed (mph)	Number	Average headway (s)
Midnight	53.69	32	37.5
Early morning and night	52.94	78	15.38
Morning peak	35.45	165	7.27
Noon to evening	42.07	98	12.24

Table 4.
CORSIM simulation results.

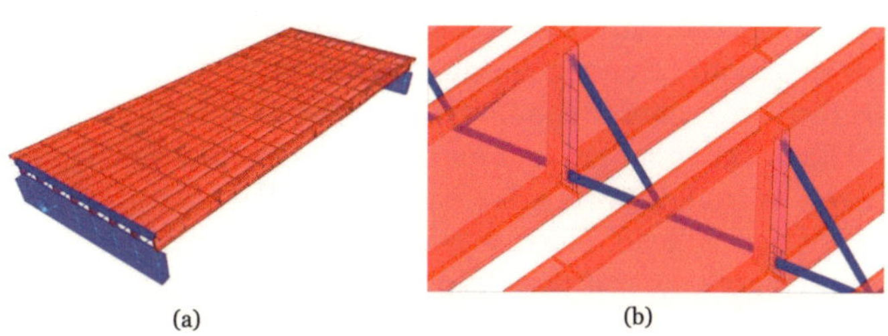

Figure 10.
Finite element model of I-270 Bridge in CSiBridge: (a) isometric view of FEM for I-270 Bridge and (b) zoom-in view (refined meshing around the welds).

embody the sectional properties of structural members (e.g., the weld between two members). In addition, fatigue damage is a local failure mode and often occurs around welded regions. Therefore, a global model with refined meshing around the welded connection between the connection plates and the bottom flanges was constructed for analysis.

3.3 Convergence test

The accuracy of a finite element analysis depends on the mesh size of the elements; the smaller the size of the elements, the greater the accuracy of the analysis. However, the desire for increased calculation accuracy can significantly increase the computational time. Therefore, in finite element analysis, a convergence test is used to determine appropriate mesh size for a model without increasing the computation time. The measurement of a finite element model's mesh size depends on the purpose of the model. Since this bridge model is to investigate the vertical stress or shear stress in the cracked connection weld, it needs to have a very fine mesh in the connection area but needs to also transit gradually to coarser meshes because otherwise the model would become unnecessarily too large. A more uniform mesh may then be used along the rest of the bridge length for all the girders. However, there are multiple parameters related to the accuracy of a two-dimensional or three-dimensional finite element model, including the dimensions and aspect ratios of the elements for the girder top flange, bottom flange, and web, as well as the bridge deck.

To simplify the convergence test for these finite element models of the I-270 Bridge over Middlebrook Road, a consistent refined mesh around the weld region was employed in all the models, and the maximum element size was used to control the uniform mesh along the bridge longitudinal length for all the girders and the deck. The determination of the first natural frequency was used as the measurement during the convergence test. As the maximum mesh size changed from 1000 in to 0.5 in, the results of the first natural frequency gradually increased from 2 to 3.20 Hz. The results of the first natural frequency were all beyond 3 Hz when the maximum mesh size of the finite element models was smaller than 200 in, which means that the error rate of the first natural frequency was under 6.25%. When the maximum mesh size was equal to or less than 50 in, the results of the first natural frequency were accurate enough with an error rate less than 2%, and were therefore used as the basis for the selection of an accurate finite element mesh in CSiBridge.

3.4 Modal analysis

Modal analysis is used to determine the vibration modes of a structure. These modes are useful to understand the behavior of the structure. They can also be used as the basis for modal superposition in response-spectrum and modal time-history load cases. An eigenvector analysis was used to determine the undamped free-vibration mode shapes and frequencies of the system.

The first six mode shapes of torsion, vertical and lateral modes are shown in **Figure 11**. To validate the finite element models, experimental data from the field test and numerical results from CSiBridge were studied. In the numerical study, the bridge was only subjected to dead load. The results obtained from the finite element model and field measurements were compared, and the differences of most of the compared frequencies were less than 6%, which was considered acceptable for the finite element analysis. All the mode shapes matched well with each other. Therefore, the CSiBridge model was considered reasonably accurate for the purposes of this study.

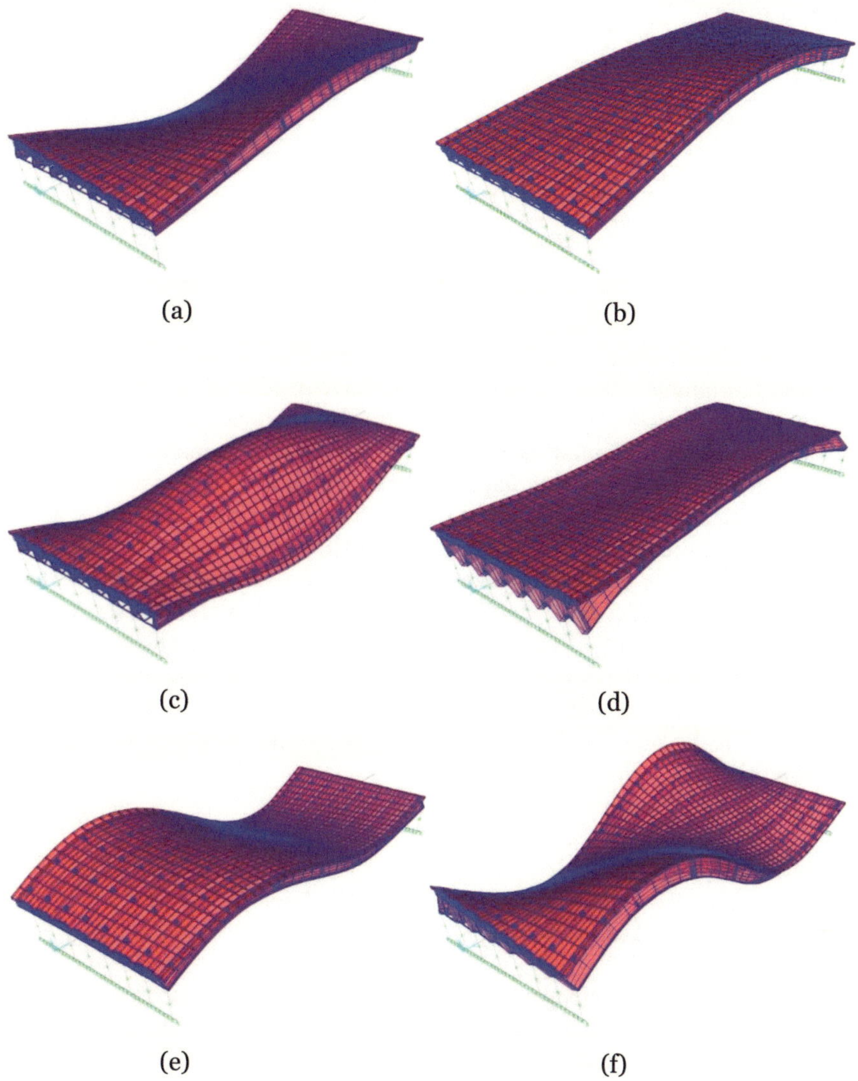

Figure 11.
Mode shapes of I-270 Bridge over Middlebrook Road in CSiBridge: (a) mode shape 1 (first torsion), (b) mode shape 2 (first vertical), (c) mode shape 3 (second torsion), (d) mode shape 4 (first lateral), (e) mode shape 5 (second vertical), and (f) mode shape 6 (third torsion).

3.5 Stresses by simulation

Possible driving forces for the fatigue cracks shown in **Figure 4** are vertical tensile stress, horizontal shear stress, or the principal tensile stress due to their combined actions along the connection welds. Live load induced stresses from the welded connections between the cross-frame connection plates and the girder bottom flanges were extracted in the refined portion of the finite element models. A total of four different traffic loading cases obtained from the traffic simulation were studied as described below and the key results were summarized in **Table 5**. For all the four cases analyzed, the longitudinal positions of trucks remained the same as for the previous deflection studies.

Time period	Max vertical stress	Max shear stress	Max principal stress
Midnight	6.665	2.165	7.629
Early morning and night	7.586	2.563	8.70
Morning peak	12.94	4.327	14.84
Noon to evening	7.905	2.664	9.061

Table 5.
Stresses in cross frame connection plate-to-girder bottom flange connections at G3 without dynamic impact (FE results) (ksi).

There were many live load cases that could have produced significant tensile stresses in the connections of concern. The simulated truck loading contained most of the possible truck loading patterns. Magnitudes of tensile stresses in the connection plates depend on the magnitudes and positions of the wheel loads of crossing vehicles. The stresses listed in **Table 5** are for illustration and are taken from the connection plates at Girder 3 for the four different time periods. A comparison of live load cases for the four different time periods suggest that live loads during morning peak may have caused the highest tensile stress of 12.94 ksi in the connection of concern. All the shear stresses in the connection welds were much lower than the vertical stresses at the same spot during each time period. Considering a factor of dynamic load allowance, the dynamic maximum vertical stress was 16.822 ksi, which perfectly matched with the field measurements.

4. Cause of fatigue cracks

4.1 Connection plate configuration

The results of the finite element analysis were verified and validated with the field test data; all the cracks were located on the western sides of the connection plates. The vertical stress near the welded edges of the connection plates followed the same pattern; the western sides of the connection plates were under tension; and the eastern sides of the connection plates were under compression. To further discuss the cause of this phenomenon, a series of controlled FEM tests were established for the comparison study.

According to the design drawings and the existing bridge construction, cross-frame connection plates and bearing stiffeners are normal to the girders and connection plates connecting the cross-frames are bent. Therefore, for the original FE model, all the connection plates were normal (90°) to the girders and the cross-frames are parallel to the two abutments with a skew angel of 76°. For the controlled model, all the connection plates were parallel to the cross frames with the same skew angel of 76 degree (**Figure 12**).

4.2 Bracing system configuration

The K-type bracing system was modeled for studying the influence of the bracing system configuration on the stress distribution in the connection plates. The K-type cross-frame without top chord was modeled in the original FE model, while the K-type cross-frame with top chord was modeled in the controlled

Figure 12.
Skewed (right) and non-skewed (left) connection plates.

model. The cross section of the diagonal and bottom chords was employed for the additional top chords (**Figure 13**). All the models were subjected to the same live load case. The live load case was defined as an HS20 truck in the right traffic lane passing across the bridge from north to south at the speed limit of 55 mph. The vertical stress at the crack location (Girder 3 Diaphragm 3) and the axial forces in the top chord located at Diaphragm 3 Bay 2, directly connecting with the crack side, were analyzed and are shown in **Table 6**. Maximum vertical stresses in the model with the non-skewed connection plates were much higher than the stresses in the model with the skewed connection plates. The maximum axial forces in the models during the load time history analysis were quite small; the values were only 3.47 and 1.12 kip. The values of maximum vertical stresses did not change much due to the addition of a top chord. It demonstrates that the connection plate configuration has a significant influence on the stress distribution in the connection plates, while

Connection plates configuration	Bracing system configuration	Maximum axial force (kip)	Maximum vertical stress of crack location (ksi)
Non-skewed connection plats	K-frame without top chord	—	13.50
	K-frame with top chord	3.47	12.66
Skewed connection plats	K-frame without top chord	—	0.33
	K-frame with top chord	1.12	0.30

Table 6.
Maximum vertical stress and axial force through simulated numerical analyses.

Figure 13.
K-frame without top chord (left) and K-frame with top chord (right).

the top chord of K-type bracing plays a negligible role in this situation. Further, the results showed that X-type or K-type bracing made no difference on the vertical stress at the crack location.

The measured high vertical tensile stress around the connection plate welds was proven caused by the configuration of the connection plates instead of the configuration of the cross-frames. The connection plates, which were bent to be parallel to the skewed abutment, induced torsion in the connection plate welds. Differential displacements between the girders caused one diagonal cross frame to be in tension and the other diagonal to be in compression. Measured vertical tensile stresses from field tests up to 16.1 ksi in the connection plate explains why fatigue cracks have occurred at their connections to the girder bottom flange. Girders 3 and 4 are located under the slow-moving lane which most heavy trucks are using while Girders 1 and 2 support a shoulder and thus larger differential deflections typically occur between Girders 2 and 3 (with up to 0.5″ to 0.75″ vertical deflections due to observed live load). The connection plate configuration is a key factor in the stress distribution that results in the connection plates.

5. Conclusions

The passage of trucks on the bridge deck can cause vertical tensile stresses in the welded connections between cross-frame connection plates and girder bottom flanges. These stresses were highest at the outer edge of the connection plate where all the existing four fatigue cracks on the I-270 Bridge over Middlebrook Road were located. Girder 4 located at the center left of the middle traffic lane, and Girder 3 located at the center right of the right traffic lane, are the most critical locations for the bridge deflections and the resulting stresses.

The live load-induced stresses in the connection plates were localized around the welded connections and would not be anticipated to spread from the bottom to the top of connection plates. At the same face of the connection plate, both tensile and compressive stresses were observed at the symmetric positions around the girder web. The cracked side of the connection plates was always under tensile stress, while the uncracked side was always under compressive stress during each time period. At the same location of the cracked side, the north face and the south face sustained the same stresses, (although opposite directions). It was proved that the high vertical tensile stress around the connection plate welds was caused by the configuration of the connection plates instead of by the configuration of the cross-frames. The connection plates, which were bent to be parallel to the skewed abutment, induced torsion in the connection plate welds. The connection plate configuration is a key factor in the stress distribution that results in the connection plates.

Different from the explicit equation-based method, the proposed approach combines a comprehensive traffic loading model, which includes information on vehicle types, axle weights, axle spacing, and the lane occupation, and a detailed 3D FE model, which enables fatigue analysis on unreachable or complicated details where complex stress conditions may exist. The proposed approach may be used as a tool accompanying a monitoring program to find the stresses in unmonitored details or to reduce the frequency of structural health monitoring resulting in lower costs in fatigue assessment. In such case, the proposed approach also provides a tool to predict the fatigue reliabilities of these hard-to-reach details. When combined with the fracture damage mechanics, the proposed approach can help understand the accumulation of fatigue damage and crack propagation.

Acknowledgements

This work was partially sponsored by the US Department of Transportation's Office of the Assistant Secretary for Research and Technology (USDOT/OST-R) under The Commercial Remote Sensing and Spatial Information (CRS&SI) Technologies Program. This support is acknowledged and greatly appreciated.

Author details

Gengwen Zhao[1], Chung C. Fu[2*], Yang Lu[3] and Timothy Saad[2]

1 Virginia Department of Transportation, Richmond, VA, USA

2 The Bridge Engineering Software and Technology (BEST) Center, Department of Civil and Environmental Engineering, University of Maryland, College Park, MD, USA

3 Asian Development Bank, Mandaluyong, Metro Manila, Philippines

*Address all correspondence to: ccfu@umd.edu

References

[1] Banjara NK, Sasma S. Evaluation of fatigue remaining life of typical steel plate girder bridge under railway loading. Structural Longevity. 2013;**10**:151-166

[2] AASHTO. Guide Specifications for Fatigue Evaluation of Existing Steel Bridges. Washington, D.C: American Association of State Highway and Transportation Officials; 1990

[3] Schilling CG. Stress cycles for fatigue design of steel bridges. Journal of Structural Engineering. 1984;**110**(6):1222-1234

[4] Raju S, Moses F, Schilling C. Reliability calibration of fatigue evaluation and design procedures. Journal of Structural Engineering. 1990;**116**(5):1356-1369

[5] Nowak AS, Nassif H, DeFrain L. Effect of truck loads on bridges. Journal of Transportation Engineering. 1993;**119**:853-867

[6] Laman JA, Nowak AS. Fatigue-load models for girder bridges. Journal of Structural Engineering. 1996;**122**:726-733

[7] Miao TJ, Chan THT. Bridge live load models from WIM data. Engineering Structures. 2002;**24**:1071-1084

[8] NCHRP. Protocols for Collecting and using Traffic Data in Bridge Design. Washington, D.C: Transportation Research Board; 2011

[9] Chen S, Jun W. Dynamic performance simulation of long-span bridge under combined loads of stochastic traffic and wind. Journal of Bridge Engineering. 2010;**15**:219-230

[10] Internet Traffic Monitoring System. Maryland Department of Transportation State Highway Administration. Available from: http://maps.roads.maryland.gov/itms_public. Accessed Jul. 29, 2015

[11] ITT Industries, Inc., Systems Division: ATMS R&D and Systems Engineering Program Team; Colorado Springs, CO 80935-5012

[12] CSiBridge. Integrated 3D Bridge Design Software. Berkeley, CA: Computer and Structures, Inc; 2010